John Henry Napper Nevill

The Biology of Daily Life

John Henry Napper Nevill

The Biology of Daily Life

ISBN/EAN: 9783337095437

Printed in Europe, USA, Canada, Australia, Japan

Cover: Foto ©berggeist007 / pixelio.de

More available books at **www.hansebooks.com**

THE BIOLOGY OF DAILY LIFE.

MY PURPOSE IS SIMPLY TO SHOW THAT A RATIONAL POLICY MUST RECOGNIZE CERTAIN GENERAL TRUTHS OF BIOLOGY; AND TO INSIST THAT ONLY WHEN STUDY OF THESE GENERAL TRUTHS, AS ILLUSTRATED THROUGHOUT THE LIVING WORLD, HAS WOVEN THEM INTO THE CONCEPTIONS OF THINGS, IS THERE GAINED A STRONG CONVICTION THAT DISREGARD OF THEM MUST CAUSE ENORMOUS MISCHIEFS.

HERBERT SPENCER in *The Study of Sociology* (p. 346).

DICTAT RATIO (SI QUID EGO HIC JUDICO,) MORBUM, QUANTUMLIBET EJUS CAUSAE HUMANO CORPORI ADVERSENTUR, NIHIL ESSE ALIUD QUAM NATURAE CONAMEN, MATERIAE MORBIFICAE EXTERMINATIONEM IN AEGRI SALUTEM OMNI OPE MOLIENTIS.

TH. SYDENHAM, M.D., *Opera Omnia* (p. 26).

[REASON (IF SUCH AN ONE AS I MAY PRONOUNCE ANY JUDGMENT) REQUIRES US TO BELIEVE, THAT, BE THE SYMPTOMS OF IT, WHICH AFFLICT THE HUMAN BODY, NEVER SO SEVERE, DISEASE IS NOTHING ELSE BUT THE EFFORT OF NATURE, ATTEMPTING BY EVERY MEANS THE EXTERMINATION OF THE DISEASE-PRODUCING MATTER, FOR THE HEALTH OF THE PATIENT.]

THE BIOLOGY OF DAILY LIFE

BY

JOHN HENRY NAPPER NEVILL, M.A.

SOMETIME A STUDENT OF MEDICINE IN TRINITY COLLEGE, THE ROYAL
COLLEGE OF SURGEONS, AND THE MEATH HOSPITAL, DUBLIN

LONDON
KEGAN PAUL, TRÜBNER, TRENCH, & CO. (LIMITED)
LUDGATE HILL

1890

CONTENTS.

	PAGE.
PREFACE	vii
THE SCOPE OF BIOLOGY	1
(From Herbert Spencer's *Synthetic Philosophy*.)	

THE BIOLOGY OF DAILY LIFE.

CHAPTER I.—The Law of Interchange . . . 3

CHAPTER II.—The Law of Interchange in relation to the Body in health and its aliment . . . 8

CHAPTER III.—The Law of Interchange in relation to the Body in disease, and the remedies for disease . 15

CHAPTER IV.—How the Law of Interchange explains the relations between the Body and lower organisms, particularly micro-organisms, in health and natural decay 26

CHAPTER V.—How the Law of Interchange explains the relations between the Body and lower organisms, particularly micro-organisms, in disease and non-natural death 43

CHAPTER VI.—The Protoplasm theory particularized and tested by facts, and re-stated with the necessary corrections 64

APPENDIX TO CHAPTER VI.—Remarks on the general reasoning on which the conclusions of the specific microbe theorists are based, and on some of the methods employed by them for classifying and identifying micro-organisms 86

CHAPTER VII.—How the three desiderata of the celebrated physician Sydenham have been discovered after two hundred years of waiting 94

CHAPTER VIII.—The Banquet of Alma, or Diet of Health, not a meagre fare, but while so cheap as to render starvation almost an impossibility, when once the truth is fully known, can be even luxurious . . 118

CHAPTER IX.—The importance of using no intractable materials, for the construction of the Human Body . 133

PREFACE.

To clear the way for the understanding of the line of argument adopted in this little book, I commence with a quotation from Herbert Spencer's *Synthetic Philosophy*, on the scope of Biology.

I do this for two reasons:

1. To present the reader with the fullest and clearest explanation obtainable, of the meaning of the words "Biology" and "Biological."

2. To show what are those principles or conclusions of this science, which Herbert Spencer takes as fundamental or axiomatic, in constructing his system.

I make no use whatever of the Spencerian system in itself. I simply say at the outset of my work, I am justified in taking it as granted that such and such principles may truly be regarded as proved, verified and generally accepted, because they are a portion of those very principles which that great and accurate thinker is content, or rather constrained in constructing his Synthetic Philosophy, to take as axiomatic or fundamental, in the special area of Biology.

The foundations of the Spencerian system are

laid deep, as all the thinking world knows, in this department of science.

Even those who may find fault with the superstructure, cannot deny the reality of its deep foundations.

A few words about the origin of this little work will explain and apologize for certain personal allusions, particularly in the seventh chapter.

Last March, a German professor of chemistry wrote to me from a town in Saxony, asking for an "exhaustive and impartial" account of the system of Mr. Joseph Wallace. This set me to try and explain so far as I could (keeping carefully to the outside of the system itself) the connection between Wallace's discoveries and generally accepted scientific teaching in Chemistry and Biology.

J. H. N. NEVILL.

The Vicarage,
  Stoke Gabriel, S. Devon.
 January, 1890.

THE BIOLOGY OF DAILY LIFE.

THE SCOPE OF BIOLOGY.

FROM HERBERT SPENCER'S SYNTHETIC PHILOSOPHY.

"In the chapters treating of Organic Matter, the Actions of Forces on it, and its Reactions on Forces, the generalizations reached were these:— That organic matter is specially sensitive to surrounding agencies; that, in consequence of the extreme instability of the compounds it contains, minute disturbances can cause in it large amounts of re-distribution; and that, during the fall of its unstably-arranged atoms into stable arrangements, there are given out proportionately large amounts of motion. We saw that organic matter is so constituted that small incident actions are capable of initiating great reactions—setting up extensive structural modifications and liberating large quantities of power.

"In the chapters just concluded the changes of which Life were made up were shown to be so adjusted as to balance outer changes. And the general process of the adjustment we found resolves itself into this; that if in the environment there are any related actions, A and B, by which the organism is affected, then if A produces in the organism some change a, there follows in the organism some change

b, fitted in time, direction, and amount to meet the action B—a change which is often required to be much larger than its antecedent.

"Mark, now, the relation between these two final results. On the one hand, for the maintenance of that correspondence between inner and outer actions which constitutes Life, an organism must be susceptible to small changes from small external forces (as in sensation), and must be able to initiate large changes in opposition to large external forces (as in muscular action). On the other hand, organic matter is at once extremely sensitive to disturbing agencies of all kinds, and is capable of suddenly evolving motion in great amounts. That is to say, the constitution of organic matter specially adapts it to receive and produce the internal changes required to balance external changes.

"This being the general character of the vital Functions, and of the Matter in which they are performed, the science of Biology becomes an account of all the phenomena attendant on the performance of such Functions by such Matter—an account of all the conditions, concomitants, and consequences, under the various circumstances fallen into by living bodies."
—(*Principles of Biology*, vol. i., pp. 94 and 95.)

CHAPTER I.

THE LAW OF INTERCHANGE.

IT is a long known, well established and now universally acknowledged fact, that the mineral, vegetal, and animal kingdoms bear a definite relation to each other. Plants are intermediary as regards sustenance, between minerals and animals—a necessary link in the chain of being.

This has long been known, and is now a *commonplace*—a household word—of Biological science.

Professor Asa Gray, in his "Structural and Systematic Botany" (New York, 1862), a standard work in its day, says (page 23):

> "Plants live directly upon the mineral kingdom. *They alone convert inorganic or mineral into organic matter;* while animals originate none, but draw their whole sustenance from the organized matter which plants have thus elaborated."

In a standard medical work of the present day (Quain's *Dict. of Medicine*, 1886), Dr. Pavy writes in his article on "Aliment" (p. 81):

> "The aliment of organisms belonging to the vegetable class is derived from the inorganic kingdom. Under the influence of the sun's rays the inorganic principles are applied to growth, and constructed into organic com-

pounds. This constitutes the main operation of vegetable life, and in it we have the source of the aliment of animals, which can *only* appropriate *organic compounds*, and which, either directly or indirectly, derive these compounds from the vegetable kingdom.

"As the solar force, employed in the construction of organic compounds through the agency of the vegetable organisms, becomes locked up in the compound formed, such compound represents matter combined with a definite amount of latent force. In the employment, therefore, of organic matter as aliment by animals we have to look upon it, not only as yielding the material required for the construction and maintenance of the body, but as containing and supplying the force which is evolved under various forms by the operations of animal life."

The position of plants, as a needful link, between the mineral and animal, in the matter of food supply, is, as we see, well known.

To complete the picture it is only necessary to remember the *economy* of Nature, in using over and over again the same materials. Think of the vast army of scavengers, belonging both to the animal and vegetable worlds, which are employed to use up every decomposing particle of vegetable or animal substance, and hasten its resolution into the inorganic state, as water, carbonic acid, ammonia and such like.*

* See "*Fungi*," by Cooke & Berkeley. (Internat. Scient. Series, p. 222.)

We find, as a matter of observation, that the sustenance of those plants which are suited for the food of animals, is itself derived (not from the mineral world *in general*), but most chiefly from those mineral particles which had previously been incorporated into organisms.

This completes the picture. We see the mutual dependence of the three kingdoms, as regards supply upon each other. It is in fact *this very question of supply* which most clearly marks the boundaries of *those* Three Kingdoms, united and yet distinct, which we call animal, vegetal, and mineral kingdoms. This mutual dependence we shall name the "LAW OF INTERCHANGE." But in the following chapters we shall only need to keep in mind that well established and generally acknowledged portion of this Law, which declares that, ANIMALS DRAW DIRECTLY OR INDIRECTLY THEIR WHOLE SUSTENANCE FROM THE ORGANIZED MATTER WHICH PLANTS HAVE ELABORATED.

While this *Law of Interchange* may be well said to be generally acknowledged, yet some more or less obvious deductions or corollaries, consequent upon this law, are either not seen as they ought to be, or are practically disregarded.* Let us consider some of these in the following chapters.

---

* Any reader may see for himself a remarkable example of this practical disregard of a principle, which at the same time is fully stated in words in this very article on "ALIMENT" in Quain's Dictionary. We have seen it stated (to put the sense of the above quotation shortly): "We have to look upon the organic matter employed as aliment for animals, not only as yielding the *material*, but also as supplying the *force*, &c. This force-supply being derived *from the latent solar force locked up in the com-*

pound; formed through vegetable agency." A main question we should expect in determining the goodness of aliments would be, What kind of food supplies this *solar-plant* force, in *fullest* force and purest form? But no, in the remainder of the article this solar force is quite left out of the reckoning. The constituents of food are viewed, as the writer says, "from a *scientific standpoint*." He gives an interesting and learned account of different classifications of food—such as Prout's and Liebig's. He shows clearly that both Prout and Liebig fail, as correct interpreters, of Nature's arrangements, and then proceeds to recommend as "practically convenient" a grouping of alimentary principles "*based on chemistry.*" Indeed throughout, the standpoint is *chemical and not biological.* The result is well seen in a subsequent article, by the same writer on "DIET." He says: "The required principles are contained in food derived from both the animal and vegetable kingdoms, and the diet may be drawn from either; but looking to man's general inclination, and the conformation of his digestive apparatus, it *may be assumed* that a mixed diet is that which is designed in the plan of Nature for his subsistence" [a queer way of getting at Nature's plan], "and it is that upon which he attains the highest state of physical development and intellectual vigour." "Animal food being identical in composition [*i.e.*, from *chemical* standpoint] with the body to be nourished by it, is in a state to be more easily appropriated than vegetable food. It also appeases hunger more thoroughly, and satisfies longer; in other words, it gives, as general experience will confirm, greater stay to the stomach ("Quain's Dict.," p. 361) [say rather, it *stays in* the stomach so much longer].

But where, and O where, is my "*solar force locked up in the compound, by vegetable agency,*" gone? Like poor Pickwick, who was wheeled into the "pound" and forgotten, until his genial presence was sadly missed, the solar force is left in the compound, unthought of.

If animal flesh *is* such excellent food, and mineral salt such a valuable "inorganic constituent" of food, why do they treat their friends *so scurvily!* Chemistry can give no explanation, *but Biology* can. Name a disease that springs from a too exclusively vegetable diet! You cannot, there is none. But we all know, or ought to know, the effect of living exclusively on even the best salt flesh-meat.

SCURVY, the plague of the dead flesh and mineral salt eater and its cure by *lemons*, purest products of sun and air and plant-power, ought to convince any thoughtful mind, capable of estimating *tendency*, and not waiting to learn like the fool by bitter experience, that the plant-prepared food is best for man.

CHAPTER II.

THE LAW OF INTERCHANGE, IN RELATION TO THE BODY IN HEALTH, AND ITS ALIMENT.

SUBSTANCES belonging to the mineral kingdom (such as carbon, oxygen, hydrogen, nitrogen, phosphorus), are the ultimate constituents of all animal bodies, and therefore of our human body. This we know and confess. We go further. We confess that we are partly made of metals; iron, sodium, potassium, calcium, and perhaps other metals, are *naturally* in our bodily structure—part of our true texture.

We may go still further. Without being hypochondriacs (like that poor lady who believed she was her own silver tea-pot which she had loved as herself), we may well believe that we are, perhaps, in some sense *altogether metallic!* Since oxygen, hydrogen, and nitrogen *may* (as advanced chemists teach us) be metals in a permanent state of gaseity, I presume their compounds may be called, if we please, *alloys.* It might be gratifying to regard ourselves or to look upon our neighbours as (like those curious old heathenish bronze images of the Deity) metallic alloys!

The thing I want to bring out clearly by all these statements is, *that in a chemical point of view* we may dismiss the arbitrary distinction between mineral and non-mineral, and even, except as a convenient arrangement for dividing the study of chemistry, that between Inorganic and Organic Chemistry. *This distinction belongs properly to Biology*, and not to Chemistry.*

In Life and the science of Life, which is Biology, there is no more essential distinction than that which lies between the mineral as such, and the organism, whether animal or vegetable ; between organized and non-organized material; organized and unorganized structure.

But next to this in importance comes the distinction amongst organisms themselves, based upon their mode of assimilation.

I venture to call this distinction more essential than any based upon differentiation, though differentiation covers all the difference (biologically speaking) between a Man and an Amœba. Assimilation, or the power of drawing into, and making a part of its own intimate structure, suitable portions of its environment, is *the very most essential and vital endowment* any organism can possess. Its very existence depends upon the power of assimilation. Now, I repeat, a distinction, based upon marked difference in the mode of assimilation, *must* be a most important distinction. Such is the distinction between the vegetable organism and the animal organism which we have already described under the *Law of Interchange* (see page 5).

Our corollary is this :

NO SUBSTANCE CAN BE RECEIVED INTO THE ANIMAL

---

* See Sir H. E. Roscoe, in his "Elementary Chemistry" (p. 265). "Organic Chemistry" is defined as "the chemistry of the carbon compounds," and he denies that there is any real difference between the laws of Inorganic and Organic chemistry.

BODY, AS A PORTION OF ITS TRUE STRUCTURE, IF PRESENTED TO THAT BODY IN AN UNORGANIZED STATE; AND FURTHER, THE MORE CLOSELY THE FOOD (AS CANDIDATE FOR ASSIMILATION INTO THE ANIMAL ORGANISM) CAN BE PRESENTED TO THAT ANIMAL ORGANISM IN THAT EXACT STATE OF ORGANIZATION IN WHICH IT EXISTED IN THE VEGETABLE STRUCTURE, THE MORE FULLY WILL THE LAW OF INTERCHANGE BE FULFILLED.*

All chemical processes disorganize the tissues of plants or animals. They tend to bring them nearer the condition of inorganic matter. Chemical treatment, whether in the kitchen, the store, or the laboratory, draws the food further away from that state in which it existed in the vegetable; and by so doing renders it so much less fit for the aliment of the

---

* I say state of organization, not *mechanical* state. A vegetable structure may be squeezed, pounded, pulped, or pulverized, stewed or baked, only *not incinerated*, and its intimate molecular structure, as an organized aliment, be untouched.

In the above corollary we leave out of consideration the absorption of oxygen in respiration. We are dealing only with food. We can never be sufficiently on our guard against theories which tend to obliterate or obscure natural distinctions. For instance, the protoplasmic theory, while it emphasizes the distinction between the mere mineral and the organism, tends to obliterate the no less important distinction between organisms themselves, into vegetal and animal—protoplasm belonging apparently to both vegetable and animal kingdoms. Again, the *chemical* method of appraising the value of foods and chemical classifications of food, have resulted in water being classed along with various saline matters as "*Inorganic principles of food.*" (See "ALIMENT" quoted above.) Now water (*i.e.*, as such, not worked up into an organized compound) is not *food* at all, it is DRINK, the solvent, or "vehicle," and in fact "maid of all work" as well as chief factor, in *both vegetable and animal* structures.

animal organism, and pre-eminently, for that paragon of animal organisms, the *human* body.

It might perhaps be reasonably objected, that the *logical* conclusion from this corollary would be, that *uncooked* vegetables and fruits ought then to be our proper food. This is, however, certainly not the *biological* conclusion; at least under our present conditions of life.

We ought to know that all attempts to *dictate* to Nature, and expect her to conform to our reasoning, are utterly futile. Our reasonings and deductions must be constantly tested and corrected by actual experience: they then become valuable guides to teach us how to interpret the phenomena presented to us spontaneously by Nature, or to show us *where* we can usefully experiment.

I think we can very simply and satisfactorily explain the seeming difference between the (apparent) *logical* and *biological* result, in this question of cooking. We all know there is a difference between good and bad cooking. Our digestions painfully tell that there is such a difference. Judging by the health-standard, and not by those of the pampered palate and vitiated taste, we simply say THAT cookery is *good*, which makes the food more easily digestible and assimilable, and THAT is *bad* which does the reverse.

See, if we cannot express this in relation to the *solar force*, lying latent for our use [see p. 5, note], comfortably locked up in the compounds formed solely, as well as *solarly*, for us poor animals, by *vegetable agency*.

We have seen that the biological conditions of

food, good for us, are two: (1st) Vegetable condition of organization, and (2nd) Latent solar force, or to put it shortly, "PLANT-FORM" and "SUN-POWER." Now, if any process alter the *plant-form*, it is *bad* cookery, if it diminishes the sun-power —it is *worse*, for it destroys the most vital portion of the aliment. But if I can find a process which, without really changing the *plant-form*, will at the same time *add to* the *sun-power*, then I have a process which does not destroy but does *fulfil* Nature's Law more completely; and if, in addition to this, we also mechanically save undue labour to the organism by presenting its food to it in a state easy to be dissolved, such a process is as perfect as we can imagine. Now all wholesome food and good cooking answer to this description. If we remember that all our fuels—coal, petroleum, gas, peat, wood —are really *stored up solar force*—"*bottled sunlight*," so to speak—non-chemical cookery* simply adds to the solar force in the vegetable tissues which, by the Law of Interchange, form the best aliment of man.†

---

* I say *non-chemical cookery*, to exclude dressing with mineral salt, baking-soda, vinegar, &c., and to imply *mild cookery*; heat applied by means of water, steam, olive oil, and such like.

† An eminent American professor of chemistry, Josiah P. Cooke, Jr., thus accurately and readily describes the facts mentioned here:

"All carbonaceous materials used as fuel, whether wood, coal, oil, or gas, if not themselves visibly organized, were derived from organized structures, chiefly plants; and all the light, all the heat, all the power, which they are capable of yielding, were stored away during the process of vegetable growth. The origin of all this energy is the sun, and it is brought to the earth by the sun's rays." . . .

"How it comes, how there can be so much power in the gentle

Thus the grand "*Law of Interchange*" is justified in its followers.

For man in maturity and health *that* food is best influences of the sunbeam, is one of the great mysteries of Nature. We believe that the effect is in some way connected with the molecular structure of matter; but our theories are, as yet, unable to cope with the subject. That the power comes from the sun, we know; and, moreover, we are able to put our finger on the exact spot where the mysterious action takes place, and where the energy is stored; and that spot, singular as it may appear, is the delicate leaf of a plant.

"This same carbonic dioxide, on which we are experimenting, is the food of the plant, and indeed the chief article of its diet. The plant absorbs the gas from the air, into which it is constantly being poured from our chimneys and lungs, and the sun's rays acting upon the green parts of the leaf, decompose it. The oxygen it contains is restored to the atmosphere, while the carbon remains in the leaf to form the structure of the growing plant. This change may be represented thus:

$$CO_2 = C + O=O$$
Carbonic Dioxide.   Carbon.   Oxygen.

"Now to tear apart the oxygen atoms from the carbon, requires the expenditure of a great amount of energy, and that energy remains latent until the wood is burned; and then, when the carbon atoms again unite with oxygen, the energy reappears undiminished in the heat and light, which radiate from the glowing embers. Just as when a clock is wound up, the energy which is expended in raising the weight reappears when the weight falls; so the energy, which is expended by the sun in pulling apart the oxygen and carbon atoms, reappears when those atoms again unite. . . .

"It is one of the greatest achievements of modern science, that it has been able to measure this energy in the terms of our common mechanical unit, the foot-pound; and we know that the energy exerted by the sun, and rendered latent in each pound of carbon, which is laid away in the growing wood, would be adequate to raise a weight of five thousand tons one foot."— "The New Chemistry" (*Internat. Scientific Series*, vol. ix., p. 156).

which is most purely vegetable. Flesh of animals, though of course it *is* organized material, is yet at least one degree removed from purely vegetable organization, and is therefore by so much *past the stage* at which it is best suited for our food. The flesh has already had an animal life *lived in it;* it is so far on its way round *viâ* decomposition into inorganic matter, to be organized anew, by plant-power, and thus be ready to play its part, in the composition of fresh vegetable and animal structures, in the circuit of Life.

CHAPTER III.

THE LAW OF INTERCHANGE IN RELATION TO THE BODY IN DISEASE, AND TO THE REMEDIES FOR DISEASE.

UNLESS we fancied we could, at once, improve the very ground-plan of Nature, we should, I presume, never dream of putting into our body, *with the intention that* they should remain in it, any substances (such as mercury, arsenic, and most mineral and vegetable drugs) which are not normal constituents of a healthy human body, or resolvable into such constituents.

To do this would be like mending a stone wall with bricks; it might possibly be an *improvement*, but certainly would not be a *restoration*, and the aim of all healing is professedly RESTORATION.

No thinking man would take any mineral or vegetable drug, unless it were under the belief that, after the medicine had done some temporary work in the system, it passed out of it altogether. If this were not so the body would manifestly be permanently altered, and so *deranged*, by the intended healing.

In this view the proper practice of medicine may be regarded as *a kind of surgery*, operating by almost invisible instruments. Like surgery, it is a *forcible interference with the body, for the purpose of setting right some injury, or remedying some wrong state of things.* But the surgeon never leaves his instruments permanently located in his patient! He even seldom leaves any foreign material as a permanent part of the structure. If the constructive surgeon does this, he never imagines he has RESTORED; the surgically-produced tooth, or eye, or palate, or limb, are all known as artificial or false; so we speak of a false tooth, false eye, false palate, false limb.

The same holds good in medicine, if any foreign element becomes a permanent portion of the body's structure; only there is this most tremendous difference: In such surgery, *we know where* the foreign portion is located, and we have a *known advantage* in improved appearance, or improved speaking, or chewing, or locomotion. But if the *medicine* does not entirely pass away, what then? We have *artificiality* SOMEWHERE OR OTHER. Who *knows where?* With what results? This last question we can answer only too well. DANGEROUS *artificiality*, in the intimate structure of brain, or heart, or lungs, or other vital organs, resulting in the most incurable forms of disease.

If we attend to the full meaning of the teachings of eminent writers on medical jurisprudence, to the medical evidence in trials for poisoning,* as well as to the experience of the vast army of medical practitioners, in any standard summary,† we shall find a complete agreement upon *one point*. We may express

---

* For instance, Dr. Palmer's case, and recently the Maybrick trial.

† See Dr. Rawdon Macnamara's "Medicines: their uses and modes of administration," seventh edit. (p. 858). "Habit powerfully influences the dose we should direct." By habit, the writer clearly means habit of taking the same or similar medicine, as he gives a remarkable instance of "the combined influence of disease and habit, in establishing a tolerance of otherwise potent medicines," in a case related by Zeviani, of a woman named

this in the short maxim, **All medicines are drugs;** or, more accurately, *All the mineral and most of the vegetable medicines in common use are veritable drugs,* i.e., *they do not pass entirely out of the system.* Not only does the continued use of such medicine *produce an altered habit* of the system (this alteration evidenced by increased toleration, and necessity of increasing the dose to produce equal effect), but this habit is (as, indeed, the etymology of the word "habit" implies*) a *having*, or holding in the system some part of the drug, which does not quite pass away.

Some portion, even of a single dose, will remain, and when the use of the drug has been long continued, a very serious amount is stowed away in some part of the organism. For a time this seems to be treated with a sort of toleration. The toleration of the human body for foreign material is a very noticeable thing. All constructive surgery proves that in certain cases alien substances (such as artificial teeth, eyes, tympanum of ear, cranial plate, and many more), especially when they serve any useful purpose, are tolerated by the system. They are not treated like the typical "foreign body" of the older pathologists and extradited by suppuration, but are natu-

---

Galvani who had recourse to opium to relieve pain, and in thirty-four years consumed *two hundred weights* of the crude drug, her daily dose at last being two hundred grains.

* "Habit is that which is held or retained, the effect of custom or frequent repetition."—*Imperial Dictionary.* Though this suits my argument, I think it a very inadequate meaning. Habit, I fancy, is rather the manner in which the whole system, or any part (Body, Soul, Spirit—*Disposition*, &c.), *holds itself.*

C

ralized and retained, as useful citizens in the body politic.*

This toleration, so wise and useful in its intention, can be, and is, systematically abused. Hence comes the seeming impunity wth which drugs can be taken. These dangerous aliens cannot be at once extradited, *for the law of toleration forbids this;* but they are dealt with as wisely as the circumstances admit. Let us see how.

A man forced to stow away heavy lumber in the rooms of a rather crazy house would naturally pile the heaviest weight on those parts of his flooring he deemed the soundest and strongest. He would also try to place the lumber where it would cause the least annoyance possible, and occasion as little obstruction as possible to the domestic economy of the household. But convenience would, *ex hypothesi*, have to be sacrificed to safe storage.

Now this is precisely what Nature does with a human body, under a course of ordinary medical treatment. The body of the patient is represented by the crazy house of our illustration, and the Materia Medica is the copious and inexhaustible supply of the matter represented by the lumber.

The drugs are stowed away with as little inter-

---

* A remarkable illustration of this was given me by a dentist, who adopts an ingenious mode of practically restoring partially decayed teeth. Fine platinum wires are passed into the nerve-holes, from which the nerves themselves, for some little distance, have been extracted. In this way a crown is securely fastened upon the duly prepared stump. The uncrowned stump, after the nerve has been removed, in time crumbles away, but, *if crowned*, and so made useful, although of course it is equally a dead substance, it is accepted, and retained as a living tooth. The writer can vouch for both these statements in his own person.

ference with vital functions as possible, but, as before, convenience cannot be principally studied, and Nature always chooses the *strongest and least sensitive parts.* So far all is plain sailing, but we soon find ourselves stranded in a most extraordinary paradox. What are the strongest and least sensitive to pain and external irritation of all the parts of the body? We must answer: *The strongest and least sensitive parts are the most essential and vital parts.* So we have what I may call the PATHOLOGICAL PARADOX (for it is certainly contrary to all our preconceived notions and opinions), that drugs are stowed away in the most essential and vital parts of the body, such as the brain, heart, liver, and osseous system.

But this tolerance has its limits. It is reached sooner or later if the use of the drug is persisted in. It is true that the continued use of a drug keeps up the state of enforced toleration, *because that continued use* prevents Nature from attempting the expulsion of the portions of the drug already pent up. Hence comes the fallacious hope of the drug-taker. Drug-taking gives relief by arresting Nature's efforts at expelling the alien diseases and drugs. For pain and sickness usually attend Nature's efforts to expel drugs or diseased matter.

Hence, I say, comes the temporary relief that has given rise to the proverb, "*Take a hair of the dog that bites you*"—a maxim that applies to all drugs as well as the alcohol or fusel oil of the drunkard.*

---

* May we not say that this "*dog that bites us*" has turned the sign of the "Greater Benefic" Jupiter, " ♃ " into the "℞" of the drug-prescriber? as Shakespeare tells us—"That's the dog's name. R is for the dog" (*Rom. and Jul.*, ii., 4).

But the "Castle of Alma" (to borrow the poet Spencer's allegorical name for man's body*) is meant to be the abode of Health; and ALMA, the fair Guardian of Health, cannot rest in peace while any enemies are within her walls.

With a fearful expenditure of vital energy, and often with unutterable agonies (notably in the case of iron, whose expulsion causes the periodic pains of neuralgia and *tic doloreux*), and *always* at the cost of some disintegration or tearing of the fabric of the body, these drugs are slowly expelled; Nature, as we have already seen, selecting her soundest and strongest organs as the channels of exit.

The facts which confirm and illustrate this are open to all attentive observers who have an opportunity for a sufficiently wide induction, and are not led astray by superficial appearances.

To express these facts in the words of a wide and most profound observer,

> "Those with the strongest chests are afflicted with lung diseases; those with the strongest digestive organs get dyspepsia; those noted for their mental brilliancy get diseases located in the head."†

All drugging is detrimental in two ways:

First, it puts foreign, and therefore disease producing material, into the texture of the body; and by so doing it also, *secondly*, strikes at the most fundamental law of organic life, viz., the law of *continuous change*. This law may be expressed thus:

---

* *Faery Queen* (Book ii., canto 9).

† "*Physianthropy, or The home-cure and eradication of disease*," p. 10.

"THE EXISTENCE OF ANY ORGANISM DEPENDS UPON ITS BEING ABLE TO MAINTAIN A PROCESS OF CHANGE, IN CONTINUOUS ADJUSTMENT WITH ITS SURROUNDINGS." (See Herbert Spencer and other writers on Biology, *passim*.)

Drugs are essentially *intractable*, and do not lend themselves to a process of change. In their mildest and least harmful forms, they obstruct and dam the river of the water of physical life. But this is at the best. No language can convey an adequate notion of the miseries which drugs (whether introduced under the guise of food, drink, or medicine) have brought upon mankind.

When they lie dormant in our system they "perplex and retard"* all its operations, and always tend to sink us Lethewards, towards the ever-open gates of death; but when in her effort to restore the body to health, Nature is struggling to expel these foes, then begins the weary labour and painful work.

It needs the imagination and the pen of a Milton adequately to state and depict the scenes which accompany the expulsion of these foes.

Indeed, in the strange correspondence between the spiritual and material worlds, the mighty poet has already in some sense described them.

Every part of the following description of the behaviour and employments of man's spiritual foes (in the second book of "Paradise Lost") corresponds with the effect of some one or other of the drugs in the process of their expulsion.

Prophetic of the practice of medicine and the delusive gains of the drug-taker, it is:

* Keats' *Ode to a nightingale*.

> "Vain wisdom all, and false philosophy;
> Yet, with a pleasing sorcery, could charm
> Pain for a while, or anguish, and excite
> Fallacious hope."

Then, in language faithfully descriptive of the physical, and the little-thought-of but far more terrible, *mental and moral* effects of drugs upon the human body, exhibited most in the process of expulsion; these foes

> "Bend
> Four ways their flying march, along the banks
> Of four infernal rivers, that disgorge
> Into the burning lake their baleful streams;
> Abhorred Styx, the flood of deadly hate;
> Sad Acheron, of sorrow black and deep;
> Cocytus, named of lamentation loud
> Heard on the rueful stream; fierce Phlegethon
> Whose waves of torrent fire inflame with rage.
> Far off from these, a slow and silent stream
> Lethe, the river of oblivion, rolls
> Her watery labyrinth."

Every drug-taker prepares for himself a Tartarus, watered by one or all of these infernal rivers. We have grown so accustomed to disease and pain that we actually think them *natural*, and then deny that Nature is wise and kind. Or, if religiously and devoutly disposed, we commit the same blasphemy in another form, and attribute them to the will of God.*

---

* This is clearly contrary to the teaching of Christ as given in the New Testament. Disease is represented as the work of the adversary, and so, of a case of "Asthenia" we read, "whom Satan hath bound, lo, these eighteen years." While the *independence* as well as the *interdependence* of physical and moral law is plainly asserted; God—your Father "maketh his sun to rise on the evil and on the good, and sendeth rain on the just and on the unjust." (See Luke xiii., 16, and Mat. v., 45.)

It is very interesting to remark in this connection the grand

But surely creative love and wisdom makes the human body, that Garden of the soul, a Paradise watered by a fourfold River of Life, and it is man's daily disobedience to the laws of Life that plunges him into this fearful ANTI-PARADISE,

> Τόσσον ἔνερθ᾽ Ἀΐδεω ὅσον οὐρανός ἐστ᾽ ἀπὸ γαίης.*

But to leave poetry and come to plain prose. Let any one take the trouble to ascertain the effects on the body, and through the body on the temper and disposition of the drugs in common use, and then say if this is an overdrawn picture. Take a few almost at random. See the uncontrollable irritability which attends the exhibition of iron and quinine, especially when their use is given up for a time, and the work of expulsion begins. Unutterable depression surely follows in the *wake*, or rather the seductive *sleep* of Chloral Hydrate. It plunged an accomplished poet and artist into more than Dantean gloom, and thousands make their minds unutterably sad by its use. Fusel oil, the cause of delirium tremens and countless suicides, besides all the crimes and misery which follow strong drink. Morphinism, the name lately coined to express a state "of sorrow black and deep," from even the hypodermic use of morphia, to say nothing of the

---

step in the right direction made by the illustrious Sydenham, in attributing *lasting* or *chronic* diseases to our own causation; though he still, not freeing himself entirely from the trammels of his age, attributes acute diseases to the "act of God," "Acutos dico, qui ut plurimum DEUM habent authorem, sicut chronici ipsos nos." See Sydenham's *Opera Omnia* (p. 344).

* Homer's *Iliad* (Book viii., 6). Tartaros, in the Iliad, denotes a place (shut in by iron gates and with a brazen floor) "*as far below Hades as heaven is from the earth.*"

long-established horrors of opium; loss of memory, from the use of bromide of ammonium, and such Lethean potions; the disintegration and rotting of the bones from mercury, and the special *necrosis* of the lower jaw from phosphorus, and, to mention smaller evils, the many sharp pangs of toothache, which, all unsuspected, follow the use of the seemingly harmless phosphates given as tonics, or as so-called " Chemical food."

But we must get on.

In the category of drugs we are obliged to include not only all the medicines in ordinary use, but also all substances (such as mineral salt and baking-soda) added to our food, and not in an already organized condition; besides these, many of the condiments which, though of vegetable composition, were either originally unfit for human food, or are rendered unfit by the treatment and adulteration to which they have been subjected.*

One special and invariable characteristic of a true medicine is, that it *must not be a drug*. It must be completely eliminated from the system in a day or two at the most. It must *prove* that it does so, BY SETTING UP NO HABIT. This simple test one can use and apply *on* and *for* himself. The doses of a true medicine become if anything *smaller*, a greater effect is accomplished by a diminished dose, as the condition which

---

* We are thankful, however (see Chapter viii.), that we are delivered from many absurd restrictions, founded on the mistaking the active manifestation of Nature's efforts after recovery for symptoms of disease. We are left everything that a natural taste pronounces " good."

called for its employment is rapidly passing away under its influence.

We must invert the proverb, and in regard to both food and medicine say, "*Meddle not with them that are* NOT *given to change.*"

CHAPTER IV.

HOW THE LAW OF INTERCHANGE EXPLAINS THE RELATIONS BETWEEN THE BODY AND LOWER ORGANISMS, PARTICULARLY MICRO-ORGANISMS, IN HEALTH AND NATURAL DECAY.

We have already seen enough to account for a large part of the diseases which afflict humanity, in the violation of one of the fundamental laws of well-being; a violation which has been practised long ago, and handed on from generation to generation, and, amidst all changes, unhappily maintained.

There is another law, or rather a *further statement* of this fundamental law of Biology, upon which we have not yet touched, and which we shall find is, if possible, more closely connected with our physical well-being than any we have yet considered. This may be stated thus:

THE EXISTENCE OF ANY ORGANISM DEPENDS UPON ITS BEING ABLE TO MAINTAIN A PROCESS OF CHANGE, IN CONTINUOUS ADJUSTMENT WITH ITS SURROUNDINGS, AND TO DO SO IT MUST MAINTAIN A STRUGGLE AGAINST OTHER ORGANISMS.

These opponents in the battle of Life may be either of the same or of different kingdoms; vegetable *versus* vegetable, animal *versus* animal, or vegetable *versus* animal.

The rule appears to be invariable that a higher, that is, a more differentiated organism, is liable to be attacked by hosts of organisms lower in the scale of differentiation.

We have already, in Chapter I., slightly alluded to the part which microscopic members of the organic kingdoms play in maintaining the Law of Interchange. Every decomposing particle of vegetable or animal structure is seized upon by these ubiquitous scavengers.

To bury the dead is regarded as a corporal work of mercy; Nature makes ample provision for the performance of this merciful work. As it is literally true that " in the midst of Life we are in Death," we cannot properly attend to the processes of Life without studying Nature's Burial Service. At first sight one might suppose there was no burial service, or one performed *sans cérémonie*, like Hood's pauper's funeral; but a closer view will force us to change our opinion. *Our* undertakers and gravediggers have, we hope, their moments of relaxation, but *Nature's* army of undertakers know no other trade, have no other occupation, they attend continually to this one thing. They are invisible to the natural eyes, but with the aid of the microscope we can *at least* see their *wands of office*. Like other conductors, their *leading characteristic* is the *báton*, or we may compare them to the official messengers of the House of Lords, and call them *Ushers of the Black Rod*. Nothing can be more certain than that Dame Nature spares not the rod in her school, though she never wishes to hurt with it, only to usher.

It is by the " *rod*" and the " *staff*," or more learnedly the BACILLUS and the BACTERIUM, that she ushers out her *finished* productions from one kingdom into the next. I mean from the animal into the mineral kingdom, and so on; for the mineral is the landing-stage of all organic life. I mean as regard material particles.

Let me describe them first in popular language.

To see them your microscope must magnify about 1000 diameters at least.

The Latin-named "BACCILLI" are little rod-like bodies, you distinguish them by their being more than twice as long as they are broad.

The Greek-named "BACTERIA" differ little in size, but are stouter, always shorter than double their breadth.

Like the canes in the hands of schoolmasters of the old *régime*, these rods have a habit of *splitting*—only their split or *fission* is right across, and in the middle, and they do it of themselves; thus they get as far as "twice one is two" in the multiplication table, yet at the very same stage, like "infant prodigies" of *calculation*, they go in for higher mathematics and multiply to an amazing extent, by *spores*. These spores are now believed to be identical with a group called MICROCOCCI (or *little berries*) from their appearance. These MICROCOCCI are the nearest *practical* attempt that Nature has made towards *coming to the* (mathematical) *point;* "rarely exceeding $\frac{1}{25000}$ of an inch in diameter, and often being much smaller than this."* We have just to add a sort of charnel-house or Morgue-brigade to this company, and our Lilliputian army-list is practically complete. These are the "PTOMAÏNES," or "CADAVERIC ALKALOIDS." But it may be suspected that these are semi-chemical, and not strictly a portion of the *living army that waits and lives on Death*.

* "It is now known that rod-bacteria when cultivated can produce spores which can divide and sub-divide again, and which, in their physical character, are undistinguished from micrococci." "*Quain's Dictionary*," p. 975.

Let us take the names BACCILLI, BACTERIA, MICROCOCCI as summing up the chief divisions of the vast host, utterly beyond the powers of human arithmetic to reckon or even estimate, which the microscope reveals to us.

Their chief characteristic is this. They are ever ready to rush into, and swarm in, every fluid or tissue of the higher animal and human organisms, the very moment the proper vitality of that organism is withdrawn, even partially, and when it is wholly withdrawn by death, they "take over the entire concern in a *going* condition."

I shall, I think, *prove* to the satisfaction of every candid mind that as Nature makes them, nothing can be more useful and honourable than the character and conduct of these microbes. But certain persons have made a regular business of "*cultivating*" these little plants, and then made a great name for themselves in the Scientific Market-place. They represent the microbes as the *causes* of all sorts of horrible diseases, and try to show that by a particular use of their own cultivated microbe, the disease-producing micro-organisms can be starved out.

As the popular scientific atmosphere is teeming with these representations,* we must proceed in an orderly manner to clear away the libels, first by showing how the very promptness and celerity of the little creatures have laid them open to these accusations.

I venture upon three illustrations to show how liable persons are to be misrepresented in analogous circumstances.

* See, for instance, Trouessart's "*Microbes, Ferments and Moulds.*" (Internat. Scient. Series.)

If an inhabitant of some other sphere, ignorant of the ways of men, were to come and see how even the most respectable and solemn undertaker conducts his business, he might easily fall into a serious mistake as he watched him at the task of putting away a corpse. How much more if this visitor were to come upon some plague-stricken city, when the dead-cart was going its rounds, and the dead bodies hurried out for burial. Might not such a visitor very naturally mistake these prompt and useful ministers of health and public safety for, perhaps, murderers and manslayers?

A traveller, ignorant of the habits of vultures, and unable to explain the marvellous instinct or quickness of sight or scent, which enables these aerial undertakers to be in at the death, or before death has quite come, might naturally suppose that the vultures *caused* the death of the corpse they were devouring.

But these evil surmisings would be quite erroneous. They are not the *causes*, but the *concomitants* of Death, nay rather, they are ministers of Health.

Now can we wonder that microscopists, peering down upon fragments of that wondrous parasitic world, should fall into a similar mistake; and because these pigmy hosts are ever waiting at all the avenues and doors, to seize upon the products of disease, and tread so very close upon the footsteps of retiring vitality, suppose them to *cause* what at most they only facilitate?

There is a story told* of a man who was caught in the act of bending over a victim of murder in a bed-

---

* In a pamphlet published a good many years ago, "Vacation Thoughts on Capital Punishment," by a barrister. I cite the story from memory.

room at night, with a bloody knife in his hand. The man confessed that he had come into the room, Macbeth-like, for the very purpose of committing that murder, and the knife he had clutched was the instrument by which the fatal wound had been given. Yet another man was the actual perpetrator of the crime, one who had just done the deed, and, being startled by approaching footsteps, had fled.

The would-be murderer had approached the bed with the full intention of committing that crime, when, in horrified bewilderment at finding his secret thought *acted out* before his eyes, he clutched at no air-drawn dagger, but the real weapon the first murderer had flung down. That strange tale was *true,* and proved to be so beyond all doubt. Yes; detected in the chamber of crime, in the slayer's very attitude, holding the actual instrument of death, and (for astonishment, had fixed and not quenched their expression) looking with murder-meaning eyes upon the dying victim, that man was yet only an accessory to the scene—the *concomitant*, and *not the* CAUSE, of the victim's death.

If appearances are so deceptive in the familiar world of our fellow men, can we wonder if they often lead us astray in regions so remote from our ordinary perceptions as those of which the most powerful microscopes can only give us glimpses?

Let us consider the three illustrations again :—The visitor from another sphere, the traveller, and the witnesses who came upon the would-be murderer. How should we set about to bring the truth to light in each of these cases?

(1.) To the visitor we should point out the extreme utility, nay absolute necessity, in a sanitary point of view, of quickly and expeditiously procuring the removal of the corpses, and that those who seemed to be ministers of death, were in truth a most valuable part of the arrangements for life and health.

(2.) For the scientific traveller, his fuller knowledge of the habits of vultures corrects the false impression produced by the amazing promptness with which they *almost* anticipate death, in devouring the slain.*

(3.) While the bringing the guilt to the actual perpetrator, *completely and absolutely cleared the otherwise reasonably suspected person.*

In the remaining part of the chapter I shall show how the first two of these considerations are directly applicable to the hordes of those micro-organisms which wait upon disease and death.

In the next chapter I shall endeavour to bring home the guilt to the real offenders.

To understand Disease, in any case, we must go back to its beginning, note the point of departure from the normal or healthy state. In fact, we must leave disease for the moment and go study the healthy state, which is in closest connection with that diseased state; go to the physiology to understand the pathology.

To see, then, the exact nature of the part which these micro-organisms play in diseased conditions, let

---

* Though the *Vulturidæ* are classed as a family belonging to the order *Accipitres*, which includes most birds of prey, it is acknowledged that vultures seldom if ever kill, but feed chiefly on carrion. Two species of the vulture family are actually protected by law in the Southern States of America, for their usefulness as scavengers; these are the turkey buzzard and carrion crow.

us begin by taking a summary view of the whole process, by which animal life is carried on, under normal conditions.

Let us take Herbert Spencer's own words, and follow his comprehensive and masterly presentment of the facts as attentively as we can. Then return to this subject with clear ideas.

"In the two fundamental functions of nutrition and respiration we have the means by which the supply of materials for this active molecular re-arrangement [which organisms, and especially animal organisms, display] is maintained.

"The process of animal nutrition consists in the absorption partly of those complex substances, which are thus highly capable of being chemically altered, and partly in the absorption of simpler substances capable of chemically altering them. . . .

"The inorganic substance, however, on which mainly depend these metamorphoses in organic matter is not swallowed along with the solid and liquid food, but is absorbed from the surrounding medium—air or water, as the case may be. Whether the oxygen taken in either as by the lowest animals, through the general surface, or, as by the higher animals, through respiratory organs, is the immediate cause of those molecular changes that are ever going on throughout the living tissues; or whether the oxygen playing the part of scavenger merely aids these changes by carrying away the products of decompositions otherwise caused; it remains equally true that these changes are maintained by its instrumentality. . . .

"In any case it holds good that the substances of which the animal body is built up enter it in a but slightly oxidized and highly unstable state; while the great mass of them leave it in a fully oxidized and stable state.

"It follows, therefore, that whatever the special changes gone through, the general process is a falling from a state of unstable chemical equilibrium to a state of stable chemical equilibrium. Whether this process be direct or indirect, the total molecular re-arrangement and the total motion given out in effecting it must be the same." (*Biology*, pp. 34 and 35.)

Just another much shorter and simpler quotation, this time from an eminent teacher of chemistry, as well as practical experimenter, in that science, and I think the matter will be clear to the general reader.

"The animal lives upon organized materials, taking up oxygen, and evolving carbonic acid and other oxidized products; the plant lives upon inorganic materials, especially carbonic acid, water, ammonia, and salts, organizing them, and evolving oxygen. The chemical function of the animal is oxidation, that of the plant reduction."*

Now taking the human body in a state of health

---

* Sir Henry E. Roscoe's "Elementary Chemistry," 1886, p. 410. Of course Sir Henry takes a purely chemical view, and very justly. He is not treating of physiology or biology, but of chemistry.

(or that which passes for health), we find that the above description very fairly represents what is constantly going on. Thus the component parts, or ultimate molecules of man's body, pass almost at once into the inorganic. One step, or two or three at the most, from the highest of organisms into the totally inorganic state, from the animal into the mineral kingdom, in perfect and accurate accordance with the Law of Interchange.

But when death comes, or even the shadow of it, in the shape of diminished vitality in any part, or throughout the whole fabric of the body, then we find a very different state of things.

Like the barbarian hosts which facilitated the decline and fall of the Roman Empire, so we "decline and *fall off*" (as Dickens' Mr. Wegg would say), not at once down to the ground (*i.e.*, into the mineral state), but breaking up, *en route*, into lesser systems.

Again, just as the philosophical historian sees in these barbarian hordes not the causes but the concomitants and facilitators of the Empire's downfall; and traces the causes elsewhere, such as to disorder in the centres of political and social life; so the philosophical biologist sees in these myriad microscopic hordes only the natural footsteps of Decay. For the very characteristic sign of natural movement is a sort of gliding continuity, the taking of very short steps, much as in the Æneid, Virgil describes how Venus, as she was moving away, bewrayed the secret of her true divinity to her son :

"pedes vestis defluxit ad imos ;
Et vera incessu patuit dea."

So Nature (our Alma Venus) reveals herself. We

recognize what is truly natural in the manner in which Life passes away. Her garment of living organization flows downward to the very lowest point of contact with the inorganic; by a kind of gliding continuity earth is organized into the body of man, and returns in like manner—" earth to earth."*

The reader is getting justly impatient of all this verbiage. Let us come to the facts, and try to put them in as few words as we can, consistently with perfect fairness of statement.

In a popular Encyclopædia under the word "Bacteria," we shall find a very fair expression of what may be called the popular view of what is now taken as science by educated persons not specially students of this science.

> "Bacteria are not only associated with various fermentative changes in fluids, but they also stand in a causal connection with various diseases."

Such "proofs" as the following are cited:

> "Koch found that if a large quantity of putrid material was introduced into animals they died very quickly. . . .

---

* Compare the following quotation, from Herschel's "Discourse on Natural Philosophy." "The travelling instances, as well as what Bacon terms frontier instances, are cases in which we are enabled to trace that general law which seems to pervade all Nature, the law as it is termed of continuity, and which is expressed in the well-known sentence, "Natura non agit per saltum." The pursuit of this law into cases where its application is not at first sight obvious, has proved a fertile source of physical discovery, and led us to the knowledge of an analogy and intimate connection of phenomena, between which, at first sight, we should never have expected to find any."

"Koch was able to cultivate these bacteria on gelatinized meat infusion, or solidified blood serum, where they grew in large quantities. . . . The minutest quantity of these bacteria, cultivated in this way, and inoculated into an animal, produced the original disease in its full virulence.

"Such experiments prove absolutely that the bacteria growing in the blood were the real cause of the disease. In a number of instances similar proof that bacteria are the cause of disease has been furnished, such as splenic fever, septicaemia in mice, chicken cholera, septicaemia in rabbits, malignant œdema in guinea-pigs, and tuberculosis in the lower animals." *

Whatever we may think of the reasoning, the statement is clear and bold enough—"*Bacteria cause disease.*"

But if from this we turn to a work intended for actual workers, and by a writer of high authority, as an experimenter, we find a very different tone. Very serious doubt is thrown upon the *causative* power of these microbes, and the exclusive influence they are assumed to have in specific disease *is positively* denied in nearly all the instances mentioned above.† I give a few quotations from the latest edition of Dr. Klein's "Micro-Organisms and Disease"—a work which may be regarded as the standard work on this subject, in

---

* "National Encyclopædia," latest edition.

† "Micro-Organisms and Disease," by E. Klein, M.D., joint lecturer on General Anatomy and Physiology at St. Bartholomew's Hospital.

English, and in which the latest collected account of the results of German, French, Italian, Swedish, and American research—in fact, a record of the work of physiological laboratories of the world—is given in a short form.

"Micrococci occur always normally in large quantities in the fluids (saliva and mucus, &c.) of the nasal and oral cavities, pharynx, larynx, and trachea. They are derived no doubt from the atmosphere. On the papillæ filiformes of the tongue they form in some cases large masses. Pasteur has inoculated rabbits with the saliva of a child that suffered from hydrophobia, and having cultivated artificially the micrococci present in this saliva, thought to have discovered that a micrococcus (*microbe spéciale*) is the cause of hydrophobia.

"That saliva of the healthy dog and of man inoculated subcutaneously into rabbits, sometimes produces death in these animals (Senator), had entirely escaped his notice, and Sternberg (Bulletin of the National Board of Health, U.S.A., Ap. 30th, 1881) has proved this in an extensive series of experiments. His own saliva proved sometimes fatal to rabbits. They die of what is called septicæmia, and Sternberg thinks it is due to the micrococci; but this is not to be considered as satisfactorily proved.

"All these micrococci stand therefore on no definite causal relation to the respective maladies, but are probably only of secondary importance." (*Klein*, pp. 68 and 69.)

I must refer the reader to Klein's own book, where he will see that the author is, as far as regard to facts will permit him, a strong *advocate for the prosecution* (so to speak) *of the microbes*. The samples which I give are then simply a few of the *admissions of the accusers of our microscopic brethren*.

If any microbe had been *proved* (by the sort of evidence considered as conclusive by these experimenters) to be *pathogenetic*, or productive of a special disease, it was the *anthrax-bacillus*. It may perhaps be taken as pathogenetic of the disease called Malignant Pustule, "a specific contagious disease, communicated to man from the disease of horned cattle, horses and sheep, &c., known as splenic fever. It is called also woolsorter's disease, from its connection with that trade."

> "But of this disease it is said, on the authority of observers, direct inoculation is rarely, perhaps never, from the living animal, usually from the carcase, affecting therefore chiefly butchers and slaughterers, &c. It may also arise from eating the flesh." (See Quain's Dictionary.)

Even of this bacillus we shall find that it is practically harmless UNLESS THERE HAS BEEN AN OPERATION PERFORMED (thought little of by these experimenters) but yet *proved to be dangerous*, and always productive of disease *of itself*, an assault upon Nature, *by which even healthy human spittle has proved fatal to rabbits*. (*Klein*, p. 69, sup. cit.) (Just as a crust of wholesome bread may be dangerous as a missile shot from a gun, or as an obstruction in the throat.) This is the

OPERATION OF INJECTION INTO THE BLOOD, by which these cruel inoculations are performed.

Closer observation seems to show that this so-called bacillus (bacillus anthrax) is nothing more than a virulent form of the TORULA, which I shall describe in the next chapter when on the subject of yeast, while Klein has the following remark (p. 156):

> "Pasteur's statement that in animals dead of anthrax and buried, the bacilli form spores, and that these spores are taken up by earth worms and carried to the surface of the soil, where they are deposited with their castings, and thus are capable of infecting animals grazing or sojourning on this soil, is not borne out by the above observations."

Of the *tubercle bacillus* we read (p. 170):

> "I cannot agree with Koch, Watson Cheyne, and others who maintain that each tubercle owes its origin to the immigration of the bacilli, for there is no difficulty in ascertaining that in human tuberculosis, tuberculosis of cattle, and in artificially-induced tuberculosis of guinea-pigs and rabbits, there are met with tubercles in various stages—young and old—in which no trace of a bacillus is to be found; whereas, in the same section, caseous tubercles may be present containing numbers of tubercle-bacilli."

Of Koch's famous "comma-bacillus" of cholera, we have the following summing-up of a long argument (p. 181):

> "From all this it follows that the choleraic

comma-bacilli are powerless to produce disease in guinea-pigs . . . ; and that a previous patho logical state of the intestine, such for instance as is produced by the injection into the peritoneal cavity [bowels] of considerable quantities of tincture of opium, enables the comma-bacilli to undergo multiplication."*

Not to weary the reader with quotations we conclude with the following very important one, taken from an article on PYAEMIA, or Blood-poisoning by *pus*. (*Quain*, p. 1311.)

"Before leaving the consideration of the pathology of pyaemia, it is necessary to allude to the connection which is supposed by some to exist between bacteria and this disease. It is said by Dr. Saunderson that a great number of microzymes are found in the blood and inflammatory exudations of animals suffering from acute infective fever, produced by inoculation of septic matter. Others (Wilks, Moxon, Goodhart) have failed to find bacteria in the blood of living cases of pyaemia, though they may be found in great numbers after death.† The Committee appointed by the Pathological Society 'to investigate the nature and causes of these infective diseases, known as

---

* It appears that the guinea-pig can stand the watery extract of opium, not the tincture. In any case the reader will recognise the old old story, equally true of the *human guinea-pig*, that the drugger give employment and means of living to the undertakers, and enables them "*to undergo multiplication.*"

† What a testimony this is to the delusive character of the evidence derived from animals under these horrible experiments.

pyaemia, septicaemia, and purulent infection, state that 'although bacteria of various forms were found in the blood in a number of cases, they could not be found in all the cases, nor were they discovered constantly in those cases where at one or other time they were present.'"
(*Trans. of Path. Soc.*, vol. xxx,. p. 44).

Though I have not attempted to place before the reader *one thousandth part* of the testimony which I have at my command, I believe enough has been placed before him to enable him to come to a verdict upon what has been presented. It can easily be seen if I have suppressed or altered, or given a biassed interpretation. I solemnly declare I have not. I believe it will be found I have fulfilled my undertaking, and the attentive reader will see :

1. That these micro-organisms are a valuable, and probably necessary, part of the arrangements of Nature, and that they promote and facilitate the salutary changes indicated by the Law of Interchange.

2. That, notwithstanding all appearances to the contrary, there is abundant evidence to acquit without a shadow of blame, or at least to make it impossible to convict those micro-organisms, which have been the most specifically and definitely accused of causing disease.*

In the next chapter we shall show the real disease-producer, again taking as evidence simply the admissions of those who hold the opposite view—or in truth hold a brief this time *for* the accused party.

* See Appendix to Chapter VI., p. 86 of this book.

CHAPTER V.

HOW THE LAW OF INTERCHANGE EXPLAINS THE RELATIONS BETWEEN THE BODY AND LOWER ORGANISMS, PARTICULARLY MICRO-ORGANISMS, IN DISEASE AND NON-NATURAL DEATH.

I HAVE undertaken the apparently hopeless task of tracking *the* disease-factor amongst the host of suspected micro-organisms.

I should have been utterly unable to fulfil my engagement, if the discovery had not been already made by a great observer and discoverer. In this chapter I shall limit myself to the writings and published statements of eminent medical men—*medical* biologists and physiologists. I shall prove that in the *admissions* of those who, as I shall show, may be regarded as the most able counsel *for* the accused, I shall have quite enough to secure a verdict and sentence of guilty, against the organism which I accuse of being the disease and death-factor.

In doing this I feel myself in the position of an officer of the law, whose unpleasant task it is to announce to an afflicted family, where the wife and mother has been foully murdered (and the members of that family are, in their distraction, vaguely suspecting neighbours, servants, friends, enemies, seen or unseen, or some half suspecting that the death was by suicide), that the husband and father, the respected head of the family, and the mainstay of the household, is himself the murderer.

With what utter incredulity and indignant denial

would such a charge be received. Until passion had cooled down, and scope was given for calm reason to exercise itself, what chance would there be of a hearing for so painful, so cruel, so apparently wanton and wicked a denunciation!

Now I charge with being the disease-factor in man that reputed portion of the human organism, which is regarded by most physiologists, and by all the readers of popularized science, as the very source and origin of the body, the structural unit, by whose protoplasmic power the whole bodily frame is formed.

This is the "white corpuscle" (so-called) "of the blood."

It will be absolutely necessary for the reader to understand fully and accurately the present state of scientific teaching in regard to the blood. I must therefore trouble him with full and long quotations from the very highest and newest physiological works—all of them from the writings of authorized and eminent teachers in the schools of medicine.

I shall respect the feelings of the general reader, whom I assume (if I may presume to hope I shall interest any) to be of either sex, and will introduce no needless or unpleasant medical detail, but above all I shall try to be accurate and full, and give ample authority for each statement, giving also the references, so that any reader can easily see for himself, whether the author is fairly quoted and interpreted.

I extract a description of the Blood from the fourth edition of "Elements of Histology," by E. Klein, M.D.:

> "Under the microscope blood appears as a transparent fluid, the *liquor sanguinis* or *plasma*,

in which float vast numbers of formed bodies, the blood corpuscles. The great majority of these are coloured : a few of them are colourless. The latter are called *white* or *colourless blood corpuscles*, or leucocytes. The former are called *red* or *coloured blood corpuscles*, or blood discs. They appear red only when seen in a thick layer; when in a single layer they appear of a yellow greenish colour, more yellow, if of arterial, more green if of venous blood.

"The proportions of plasma and blood corpuscles are sixty-four of the former and thirty-six of the latter in one hundred volumes of blood. By measurement it has been found that there are a little over five millions of blood corpuscles in each cubic milimétre ($\frac{1}{15625}$ of a cubic inch) of human blood. There appears to be in healthy human blood one white corpuscle for 600—1200 red ones. In man and mammals the relative number of blood corpuscles is greater than in birds, and in birds greater than in lower vertebrates.

"In a microscopic specimen of fresh unaltered blood, the red blood corpuscles form peculiar shorter or longer rolls, like so many coins, from becoming adherent to one another by their broad surfaces. Under various conditions—such as when isolated, or when blood is diluted with saline solution, or solutions of other salts (sulphate of sodium or magnesium)—the corpuscles lose their smooth circular outline, shrinking and becoming *crenate* [*i.e.*, notched or indented]. In a further stage of this process of shrinking they lose their discoial form, and become smaller and spherical, but beset all over their surfaces with minute processes.

"This shape is called the *horse-chestnut shape*. It is probably due to the corpuscles losing carbonic acid, as its addition brings back their discoial shape, and smooth circular outline. On abstracting the carbonic acid they return to the horse-chestnut shape. Water, acid, alcohol, ether, the electric current, and many other re-agents produce discoloration of the red corpuscles; the coloured matter—generally the combination of the blood colouring matter with globuline, known as *hæmo-globine*—becoming dissolved in the plasma. What is left of the corpuscles is called the *stroma*. . . . Discoloration of the blood corpuscles can also be observed without the addition of any re-agents, or with that of indifferent fluids, such as the aqueous humour of the eye, hydrocele fluid [the water of a kind of dropsy], etc. The number of corpuscles undergoing discoloration is, however, small."

We come to the white corpuscle. As old Polonius says, I will be brief—*i.e.*, as brief as I can. Like the same fine old character, making as few comments as I can.

" 13. The **white or colourless blood corpuscles** are in human blood of about $\frac{1}{2000}$ to $\frac{1}{2500}$ of an inch in diameter, and are spherical in the circulating blood, or in blood that has just been removed from the vessels. Their substance is transparent, granular-looking protoplasm, containing larger or smaller bright granules. These granules, though usually of a fatty nature, are in some kinds of blood, notably horses', of a reddish colour, and these corpuscles are supposed by some observers (Semmer and Alexander Schmidt) to be intermediate between red and white corpuscles. The protoplasm of the colourless corpuscles contains glycogen (Ranvier, Schäfer). In the blood of the lower vertebrates the corpuscles are much larger than in mammals. But in all cases they consist of protoplasm, include one, two, or more nuclei, and show amœboid movement. This may be observed in corpuscles without any addition to a fresh microscopic specimen of blood, but it always becomes much more pronounced on applying artificial heat of about the degree of mammals' blood. It is then seen that they throw out longer or shorter filamentous processes, which may gradually lengthen or be withdrawn, appearing again at another point of the surface.

"The corpuscle changes its position, either by a flowing movement of its protoplasm as a whole, thus rapidly creeping along the field of the microscope, or it may push out a filamentous process and shift the rest of its body into it. During this movement the corpuscle may take up granules from the surrounding fluid.

"14. The white corpuscles of the same sample of blood differ in size and aspect within considerable limits, some being half the size of others, some much paler than others. The smaller examples generally possess one nucleus occupying the greater part of the corpuscle, the larger ones usually include two, three, or even more nuclei, and show more decided amœboid movement than the others.

"Division by cleavage of the white corpuscles of the blood of the lower vertebrates has been directly observed by Klein and Ranvier."* (Klein's *Histology*, p. 13.)

I now pass to the important subject of development, again, like Polonius, promising "I will be faithful." Let me just slide in one remark, the application of which to the following quotation, also *verbatim* from Klein's *Histology*, I leave to the intelligence of my readers. Do not think I am trying to "make the worse appear the better reason." I have not the skill, even if I had the malice, of Milton's sophist, Belial. (*Par. Lost*, bk. ii.)

I know my remark has a suspicious appearance of arguing that black is white.

"All black birds are not blackbirds," otherwise a sophist might prove that a crow was one of the most musical of song birds. So let me say,

All *white corpuscles* are not WHITE CORPUSCLES.

I may also explain what is meant by development, and some of the technical terms (such as mesoblast) in the quotation which follows

---

* My purpose, viz., that of showing how even generally accepted teaching in Biology contains enough to prove all that I contend for in food and medicine, is best served by quoting from the newest and most reputed standard works, such as the "Manuals for Students of Medicine," of which Klein's "Elements of Histology" is one.

The young " swell," usually called " embryo " (from the Greek ἔμβρυον = τὸ ἐντὸς βρύον, that which swells within*) grows thus :—

In the egg, or ovum (from which *every animal* begins its life), there is this embryo. It starts in life by dividing into two, and these parts again subdivide indefinitely until a mass like a mulberry is formed. Some of this mulberry-mass forms into a layer called BLASTO-DERM (Germ-skin or film). This germinal layer soon *differentiates* into three layers called (by names meaning " upper," " middle," and " lower")

      1. EPIBLAST.
      2. MESOBLAST.
      3. HYPOBLAST.

From which are developed respectively—

    1. Skin, Brain, and Nerve centres. 2. Main tissues and organs of body, arteries and veins, muscles, nerve-cords. 3. The epithelium, or lining of digestive and respiratory tracts —in a word, the inner skin, or lining of the body.

We must now call in Dr. Klein :

" 16. **Development of Blood Corpuscles.**—At an early stage of embryonic life, when blood makes its appearance, it is a colourless fluid, containing only white corpuscles (each with a nucleus), which are derived from certain cells of the mesoblast.

" These white corpuscles change into red ones, which become flattened, and their proto-

---

* Not "*from* within," *i.e.*, its own substance, as some have translated, trying to read modern embryology into ancient Greek.

plasm gets homogeneous and of a yellowish colour. All through embryonic life new white corpuscles are transformed into red ones. In the embryo of man and mammals these red corpuscles retain their nuclei for some time, but ultimately lose them. New nucleated red corpuscles are, however, formed by division of old red corpuscles.* Such division has been observed in the adult blood of certain lower vertebrates (Peremeschko) as well as in the red marrow of mammals (Bizzozero and Torre).

"An important source for the new formation of red corpuscles in the embryo and adult is the red marrow of bones (Neumann, Bizzozero, Rindfleisch), in which numerous nucleated protoplasmic cells (marrow cells) are converted into nucleated red blood corpuscles. The protoplasm of the corpuscle becomes homogeneous and tinged with yellow, the nucleus being ultimately lost. The spleen is also assumed to be a place for the formation of red blood corpuscles. Again, it is assumed that ordinary white corpuscles are transformed into red ones, but of this there is no conclusive evidence. In all these instances the protoplasm becomes homogeneous and filled with hæmoglobin while the cell grows flattened, discoid, and the nucleus in the end disappears.

"Schäfer described intracellular (endo-

---

* I venture to question the accuracy of this, the division of red corpuscles has never been observed in the blood of man, and apparently only in the *marrow, not blood*, of these mammals.

genous) formation of red blood corpuscles at first as small hæmoglobin particles, but soon growing into red blood corpuscles, in certain cells of the subcutaneous tissue of young animals. Malassez describes the red blood corpuscles originating by a process of continued budding from the marrow cells.

"The white corpuscles appear to be derived from the lymphatic organs, whence they are carried by the lymph into the circulating blood." (*Histology*, p. 15 and 16.)

Our next quotation is taken from an article in Quain's *Dictionary*. But to help the reader duly to appreciate its high authority, I will give the opinion of a writer in the *Quarterly Review* on the *Dictionary* itself. "No dictionary of medicine so compendious, and at the same time so authoritative, has yet appeared in any language. One hundred and sixty writers contribute an immense number of articles, varying in length from a column or less to thirty pages." In fact, it is one of the very highest authorities in human language at the present day, and its contributors are the leaders in their various departments.

I give a long quotation from an article on the Blood:

"The source of the red corpuscles is of the greatest pathological importance. In the embryo the blood and blood-vessels are developed from the same elements, and thus the two structures in their physiological aspect are essentially inseparable. In fully developed blood the source of the red corpuscle is obscure; but there can be no reasonable doubt that it originates in the colourless corpuscle, and more remotely in the lymphatic glands, the spleen, and the medulla of bones; and that light is of the greatest importance in the formation of hæmoglobin. With respect to the properties and *function* of the red corpuscles, it is to be noted that the ultimate elements of hæmoglobin are carbon, nitrogen, hydrogen, oxygen, sulphur, and iron."* . . . " Most important of all its properties, hæmoglobin combines with certain gases to form definite chemical compounds; with O to form oxyhæmoglobin." . . . These compounds, and especially the oxyhæmoglobin, are exceedingly unstable." . . . " Alternate oxidation of hæmoglobin and deoxidation of oxyhæmoglobin are constantly going on within the red corpuscles of the circulating blood; and the two changes occurring in the pulmonary and systemic capillaries respectively, constitute the first great function of the blood—its oxygenating or respiratory function." . . .

" It must be clearly understood that disorders connected with the red corpuscles or respiratory elements of the body, whether in amount, composition, or circulation, directly affect the oxidation-processes only.†

" Besides its origin and its function there is a third relation of the red corpuscle to the

---

* To complete the list, phosphorus, potassium, and sodium should be added. (See Schmidt's Table in "Quain's Anatomy," p. xii.)

† I shall presently show cause for believing that the red corpuscle is directly concerned in the making of *muscle*.

organism—namely, that of its *products*. These are eliminated by the ordinary channels; the salts, which are chiefly salts of potash, being excreted by the kidneys, and the coloured material furnish the pigments of the bile and urine.

" The white or colourless corpuscles of the blood, also called leucocytes, are chiefly derived from the corpuscles of the lymph, and the cells of the lymphatic glands, which they closely resemble. By escaping through the walls of the blood vessels, they become identical with the wandering cells of tissues and pus-corpuscles, from which they are indistinguishable except by locality." " BLOOD, morbid condition of" (*Quain.* p. 116).

Now we ask the reader to continue his medical course of study, and learn what the eminent authority last quoted has to say about these corpuscles.

Like Hamlet, to continue my playful illusions. I only say—"Look here upon this picture and on this, the medical presentment of two corpuscles." We shall find some startling differences, decently veiled under the professional obscurity of medical diction. Further on in the same article we read, under " MORBID CONDITIONS OF THE RED CORPUSCLES :"

"(a.) *Polycythaemia.* Increase in number of the red corpuscles is never considerable, being generally transitory and within physiological limits; for example, in the newly born, and after meals. It has already been mentioned[*]

---

[*] The exact words of this mention are " Polyhæmia is believed to be present in *plethora*, along with relative excess of the solids, and especially of the red corpuscles."

as associated with polyhaemia in plethora. In the earlier stage of cholera the red corpuscles are relatively in excess."

"(*b.*) Oligocythaemia. Diminution in number of red corpuscles is, on the contrary, of very frequent occurrence, and of the greatest pathological importance.

"The principal circumstances under which oligo-cythaemia occurs are—(1) in anaemia, or diminution of the amount of blood as a whole from any cause, whether rapid or protracted, especially as the result of fever; the red corpuscles suffering early, seriously, and persistently, as compared with the other constituents; (2) in leucocythaemia [*i.e.*, redundance of white corpuscles]—the development of the red corpuscle being interrupted;*

---

* Is it any wonder that, like Bunyan's pilgrims, one is always saying, What meaneth this? and alas, not getting as clear answers in the House of our medical INTERPRETERS. If a red corpuscle originates in the leucocytes, of which we have been told in this article "there can be no reasonable doubt," why should its development be interrupted by the mere fact of its parent's presence? Can it be that the due regard to her age, which makes a fashionable beauty keep her children from "coming out" and circulating in her stream of society, is a deep law of Nature, and runs in our blood? But see how nobly the red corpuscle shines out in sharing our bodily fortune. "*Suffering early, seriously, and persistently,*" in our sickness. Like a true friend—

> Thus of every grief in heart,
> He with thee doth bear a part;
> These are certain signs to know,
> Faithful friend from flattering foe.

Yes, our blood is our life, and the red corpuscle its most vital constituent.

(3) in hypalbuminosis, where the red corpuscles, like other elements, suffer from want of albuminous material; and (4) in chlorosis.

A little further on in the same article we come to:

"(5) MORBID CONDITIONS OF THE WHITE CORPUSCLES. — The white corpuscles of the blood may undergo certain morbid changes both in number and appearance.

"(a) The most remarkable of these is *increase* in numbers, which may advance to such a degree that the white corpuscles become as numerous as the red. This condition is known as *leucocythaemia*, or leukaemia. Short of this, however, the proportion of white corpuscles in the blood may be appreciably increased, and to this minor condition the name of *leucocytosis* has been applied.

"Leucocytosis [remember this means only a slight redundance of white corpuscles], according to Virchow, accompanies almost unexceptionally every case of lymphatic excitement, such as inflammation, and tubercular, scrofulous, or cancerous enlargement, or swelling of the glands and allied structures— Peyer's glands, the solitary follicles, the spleen and the tonsils. Leucocytosis is distinguished from leucocythaemia by its very moderate degree; by its evanescent course; by the absence or deficiency of the red corpuscles, and by the accompanying symptoms. Leucocytosis may be appreciated even by the naked eye in the clot of drawn blood, by the presence of an irregular 'lymphatic layer' — *crusta* *lymphatica*, consisting of collections of white corpuscles, between the red clot and the buffy coat, which so frequently occurs along with it.*

"(*b*) A *diminution* in the number of white corpuscles occurs in chlorosis; and it is said, in malaria, especially during the paroxysm of fever."

I have now given a full statement of the effects of *increase* and *diminution* in the case of the "red" and "white" corpuscles, given fully and fairly in the words of a believer in the white corpuscle, a believer with so robust a faith in the creed which runs: "I believe in protoplasm as the maker of all organisms, and in the white corpuscle as the special embodiment of protoplasm which makes man," that dispensing with sight, and believing what he cannot prove, he declares "there can be no reasonable doubt that the red corpuscle is formed out of the white."

Even already, before the case against it is well

---

* N.B.—The buffy coat is itself formed by leucocytes. "When human blood is drawn in inflammatory diseases, as well as in some other conditions of the system, the red particles separate from the liquor sanguinis before coagulation, and leave the upper part of the liquid clear. In this case, however, the plasma is still mixed with the pale corpuscles, which being light accumulate at the top. On coagulation taking place in these circumstances, the upper part of the clot remains free from redness, and forms the well known "buffy coat" so apt to appear in inflammatory blood." (Quain's "Anatomy," p. 32.)

The subjoined scheme, from Quain's "Anatomy," p. 28, will make the terms used clear to the reader, and show the process of coagulation:

Liquid blood { Corpuscles . { Fibrin } clot } Coagulated
              { Liquor sanguinis { Serum }   blood.

begun, the white corpuscle has an ugly connection with *fever*, with *glandular disease*, with SCROFULA, with CANCER, in the mild increase on its normal percentage, called leucocytosis, or mild white-corpuscle disease.

Now let us turn our attention to the serious increase which occasions the disease, so-called, of leucocythaemia, only noting some important points at which we have already arrived. Many of these the reader, if he possess a tolerably good microscope, can verify for himself by direct appeal to Nature, pending this he *ought* to be satisfied with the teachings of the highest medical authorities.

Let us sum up what we have already learned.

1. The red corpuscle is shaped like a quoit (discoid), the white like a ball (spherical).

The white embryotic or fœtal blood corpuscles, which develope into the red corpuscles, are simply red corpuscles not yet fully grown and coloured. The white corpuscles that turn red are not leucocytes. No observer ever saw an ordinary white "corpuscle" or leucocyte change into a red corpuscle. "*It is assumed*," and is taken for granted to suit protoplasmic theories, BUT NEVER SEEN.

2. The red and the white comport themselves like the guinea and the one pound note in the rhyme :

"A guinea it will sink and a pound it will float,
 I'd rather have a guinea than a one pound note."

3. The white corpuscles are identical with those in matter discharged in inflammation and suppuration (pus). "It is indistinguishable, *except by locality*, from pus-corpuscles."

4. The white corpuscle, in movement and habits, is

like an *amœba*.* " It is seen to take up small granules from the surrounding fluid."

5. While increase of red corpuscles practically means increase of life and health, increase of white corpuscles is at least the harbinger, if not the cause, of disease and death.

> The white makes disease,
> But the red makes muscle ;
> I would sooner have the red
> Than the white corpuscle.†

---

* The *amœba diffluens* is a common microscopic object in stagnant water. "As it flows or glides from place to place is seen to devour and digest the materials with which it is surrounded." (Jones' "Animal Creation," p. 10.)

Some one (with a turn for inaccurate biblical quotation) told an inquiring American that "the wolf and the lamb shall lie down together." " I guess," was the shrewd reply, "that the lamb will lie to the *inside* of that wolf." It is even so with the leucocyte and its granules.

A romantic explanation is suggested in the latest edition (1889) of Huxley's and Martin's "Elementary Biology," p. 373. "Occasionally an amœba has been seen to engulf another of smaller size than itself; and there is reason to believe that this process, originally thought to have been one of cannibalism, may probably be one of conjunction of dissimilar individuals for reproduction, such as is seen in the bell-animalcule."

Amœbic marriage customs are certainly peculiar, but then so are some human marriage customs. So, though we cannot tell which is the bride and which is the bridegroom, let us wish that all may be "merry as a marriage-bell."

That the amœba of the stagnant pool is a near blood relation of ours, the following quotation from the same book seems to intimate:

"If amœbae are not to be found, their nature may be understood by the examination of the *colourless corpuscles* of the blood." (Ibid., p. 372.)

† That the "red makes muscle" is undeniable, for as a respiratory agent it helps to make all the tissues of the body. I submit

But we must return to our Dictionary of Medicine.

"LEUCOCYTHAEMIA (λευκὸς, white, κύτος, a cell, and αἷμα, blood.)

"DEFINITION.—A chronic disease, in which there is a considerable and permanent increase in the number of the pale blood corpuscles, usually associated with enlargement of the spleen, sometimes also with that of the

the following argument more as an example of keeping the several lines of *physics*, *chemistry*, and *biology* distinct (like the vertical columns in an addition sum), and "*carrying*" from one to the other without confusion (as in the calc-spar fallacy, described at p. 72), than for its intrinsic value.

1. *Physically*. The microscopic appearance of striped and unstriped muscles resembles aggregates, or rather agminations (ranked arrangements) of red corpuscles. (See per Microp. and Figures in Quain's "Anatomy," pp. 65-128.)

2. *Chemically*. The ultimate composition of flesh (dried) and blood (dried) is identical. Flesh is simply solidified blood. (See analysis of dried ox-blood and beef, in Quain's "Anatomy," p. 33.)

3. *Biologically*. The tendency of the red corpuscles to agminate or form rolls (like coins in a pile) is most likely biological, and not merely physical (as suggested in Quain, &c.) In a drop of blood drawn from a perfectly healthy person of calm temper, the tendency to "pile" is not very marked, but if one has been even mentally "put out," or is from any cause in a slightly feverish state, *then* the roll-making tendency is both marked and prompt. May not this be an anticipation or hastening of a natural (vital) process, and be received as a hint from Nature as to her mode of supplying muscular tissue?

4. *Physiologically*. The connection between muscular action and increased blood supply being so immediate and *prompt*, also the directly accelerated respiratory as well as cardiac action, may possibly be taken as indicative of a peculiarly close connection between the muscle and the κατ' ἐξοχήν *respiratory* element of the blood.

lymphatic glands, and with diseases of the medulla of bone.

"HISTORY.—Pallor of the blood, as if pus were mixed with it, was noted by Béchat in the beginning of this century; and the combination of this appearance, with enlargement of the spleen, was observed by Velpeau in 1827. The dependence of this alteration in the blood on an excess of pale corpuscles was described by Donné in 1844, and interpreted[*] by him as due to imperfect transformation of white into red corpuscles.

"In 1845 two cases of this disease were published together, the one by Dr. Craigie, the other by Dr. Hughes Bennett; and to the latter appears to belong the credit of recognising the salient features of the affection as a distinct malady. A month later, however, Virchow published another case, independently and admirably worked out.[†]

"In all these cases the changes in the blood were only recognised after death.

"It was first observed during life in 1846 by Dr. H. W. Fuller, and subsequently by Dr. Walshe. In Germany the first case was diagnosed during life by Vogel, 1848. Since then numerous cases and descriptions of the

---

[*] How often are observations correct, while the interpretation is faulty!

[†] There is ambiguity about this sentence, but we *think* it means *not* that the *patient* was either (1) worked out of life, or (2) worked into health, the latter alternative is scarcely possible, but simply that the description of the disease was admirable.

disease have been published, of which the more important are those of Virchow, Hughes Bennett, Vidal, Huss, Ehrlich, and Mosler."

Omitting what is not necessary for the general reader, and merely noting one remark under DIAGNOSIS, that "if the proportion of white corpuscles to red is greater than one to twenty, the case is certainly one of leucocythaemia."

I pass on to the prognosis, which I quote in full.

"PROGNOSIS.—The prognosis of a disease which depends on a primary affection of the blood-forming organs is necessarily most grave. No means of arresting the progress of the developed disease has yet been discovered. The immediate prognosis is less serious in proportion as the evidence of organic changes in the blood-forming organs is slight and in proportion to the early stage of the disease. Neither age, sex, nor causation afford prognostic information. The greater the number of white corpuscles and the deficiency of red, as ascertained by counting, the worse the prognosis. The size of the spleen alone affords little information. Hæmorrhages are of grave augury, but epistaxis [nose-bleeding] least so."

I submit that in the above quotations (and quotations of similar import might be given to an extent only limited by the patience of the reader), we have a clear *prima facie* case against the leucocyte, that it is an invariable accessory to disease, and that of the most serious and deadly sort.

Let me again remind the reader that the above quotations are from writers who, besides being eminent authorities themselves, give the views of leading physicians, physiologists, and specialists, every one of whom adheres to the belief that the white corpuscle is the source of the red, and, in fact, the protoplasmic structural unit of the human body.

I am well aware that to some minds this will be a strong argument in favour of the leucocyte. What good cause all these great and renowned men *must* have, to make them hold to this opinion in spite of appearances, which even a tyro can appreciate! I must remind all such objectors that Truth and Nature, like Him whose word and act they represent, are no respecters of persons. If there is any one lesson which the history of scientific progress teaches more repeatedly than any other it is this—that the highway of progress is paved with discarded theories, once useful and held in high repute, and afterwards trodden under foot. True learning's royal road is a sort of macadamized highway, and the smaller the fragments used in metalling it (short of actual pulverization) the better for the road.

I fight against no well-observed phenomena—I wage no foolish war with facts, but I contend that the time has come to lay aside a *theory* that has long ceased to square with observed facts.

In Chapter V. I shall go fully into the question. Now I simply state that it is the protoplasm hypothesis which has thus outgrown its usefulness, and I end this chapter with the weighty words of Sir John Herschel:

> "What in the actual state of our science is far more important for us to know is whether our theory truly represents *all* the facts, and

include all the laws to which observation and induction lead. A theory which did this would, no doubt, go a great way to establish any hypothesis of mechanism or of structure, which might form an essential part of it; but this is very far from being the case except in a few limited instances; and till it is so, to lay any great stress on hypotheses of the kind, except in so much as they serve as a scaffold for the erection of general laws, is to "quite mistake the scaffold for the pile." Regarded in this light hypotheses have often an eminent use; and a facility in framing them, if attended with an equal facility in laying them aside when they have served their turn, is one of the most valuable qualities a philosopher can possess; while, on the other hand, a bigoted adherence to them, or indeed to peculiar views of any kind, in opposition to the tenor of facts as they arise, is the bane of philosophy." (Herschel, *Nat. Phil.*, p. 204.)

CHAPTER VI.

THE PROTOPLASM THEORY PARTICULARIZED, AND TESTED BY FACTS, AND RE-STATED WITH THE NECESSARY CORRECTIONS.

HERSCHEL, in his "Discourse on the Study of Natural Philosophy" (Chapter VII.), lays down the three ways by which we arrive at general laws.

"We have next to consider the laws which regulate the action of these our primary agents; and these we can only arrive at in three ways:

"1st. By inductive reasoning; that is, by examining all the cases in which we know them to be exercised, inferring, as well as circumstances will permit, its amount or intensity in each particular case, and then piecing together, as it were, these *disjecta membra*, generalizing from them, and so arriving at the laws desired.

"2nd. By forming at once a bold hypothesis, particularizing the law, and trying the truth of it by following out its consequences and comparing them with facts; or,

"3rd. By a process partaking of both these, and combining the advantages of both without their defects, viz., by assuming indeed the laws we would discover, but so generally expressed that they shall include an unlimited variety of particular laws; following out the consequences of this assumption by the application of such general principles as the case admits; comparing them in succession with all the particular cases within our knowledge; and lastly, *on this comparison*, so modifying and restricting the general enunciation of our laws as to *make the results agree.*"

It is by the third of these methods that I have set to work upon the "bold hypothesis," or rather "assumption," which may be briefly named the "protoplasm theory." We shall find, I believe, that that theory must be greatly modified before it can be accepted as representing a law of Nature. I shall state it in this modified form in the course of this chapter, now I proceed to the examination of the hypothesis itself.

But before I begin, let me confess my fears. I know well I am trying to approach the dearest "Idol of the (Biological) Theatre."* I am much encouraged by the words of Dr. Burdon Saunderson, who calls† protoplasm "a worn out Deus ex machinâ," and denies that it avails to explain the phenomena of function in living organisms.

* "There is also a fourth kind [of illusions] which we denominate Idols of the Theatre, and is superadded from false theories or systems of philosophy, and erroneous laws of demonstration." (Bacon's *Advancement of Learning*, Book v., chap. 4.)

† At the Newcastle meeting of the Brit. Assoc., in Sept., 1889. "Whenever this point" [viz., when we are "face to face with functional differences which have no structural difference to explain them") "is arrived at in any investigation structure must for the moment cease to be our guide, and in general two courses or alternatives are open to us. One is to fall back on that worn out Deus ex machinâ, protoplasm, as if it afforded a

First let us understand *precisely* what protoplasm means. We are doubtless familiar with its general outline, as it looms large over the whole field of biological thought, but may not be quite sure of all its features, as it issues from the hands of its makers.

I may say that the word "Protoplasm" was first invented by Hugo von Mohl, but I think at least all English-speaking people will agree that to Professor Thomas Huxley belongs the credit of making the protoplasm theory more than a mere "working hypothesis." He has at all events popularized, if he

sufficient explanation of everything that cannot be explained otherwise, and accordingly to defer the consideration of the functions which have no demonstrable connexion with structure, as for the present beyond the scope of investigation; the other is, retaining our hold of the fundamental principle of correlation, to take the problem in reverse—*i.e.*, to use analysis of function as a guide to the ultra-microscopical analysis of structure. I need scarcely say that of these two courses the first is wrong, the second right; for in following it we still hold to the fundamental principle that living material acts by virtue of its structure, provided that we allow the term structure to be used in a sense which carries it beyond the limits of anatomical investigation—*i.e.*, beyond the knowledge which can be attained either by the scalpel or the microscope. We thus proceed from function to structure, instead of the other way. In thus changing direction we are not departing from the traditions of our science."

I cannot pretend to define the term "structure" in this transcendental sense, which carries it beyond the limits of anatomical or microscopic investigation. It presents itself to me as a sort of castle in the air. But the last sentence I can fully understand, from my recollections of a circus. If one is walking in a straightforward course, of course to right-about-face is to turn your back upon all you were once aiming at, but not so when your course is circular—a man who argues in a circle can reverse his procedure without departing from his traditions. He argues in that very circle still.

has not originated, the notion that protoplasm is the "Physical Basis of Life."

We cannot do better than quote his own words, no one has at his command clearer, more picturesque, and when the occasion calls for it, grander language.

"You are doubtless aware that the common nettle owes its stinging property to the innumerable stiff and needle-like, though exquisitely delicate, hairs which cover its surface. Each stinging-needle tapers from a broad base to a slender summit, which, though rounded at the end, is of such microscopic fineness that it readily penetrates, and breaks off in, the skin. The whole hair consists of a very delicate outer case of wood, closely applied to the inner surface of which is a layer of semi-fluid matter full of innumerable granules of extreme minuteness. This semi-fluid lining is protoplasm, which thus constitutes a kind of bag full of a limpid liquid, and roughly corresponding in form with the interior of the hair which it fills. When viewed with a sufficiently high magnifying power, the protoplasmic layer of the nettle-hair is seen to be in a condition of unceasing activity. Local contractions of the whole thickness of its substance pass slowly and gradually from point to point, and give rise to the appearance of progressive waves, just as the bending of successive stalks of corn by a breeze produces the apparent billows of a cornfield.

"But, in addition to these movements and independently of them, the granules are driven,

> in relatively rapid streams, through channels in the protoplasm which seem to have a considerable amount of persistence." (*Lay Sermons*, pp. 123 and 124.)

Here we have a life-like picture of protoplasm, drawn by a master-hand. Let us now see how it applies to the particular case of man.

Happily, I can give the words of the same master of picturesque language.

After referring to the " fact that plants can manufacture fresh protoplasm out of mineral compounds, whereas animals are obliged to procure it ready-made, and hence in the long run* depend upon plants," Professor Huxley proceeds to say:

> " With such qualifications as arise out of the last-mentioned fact, it may be truly said that the acts of all living creatures are fundamentally one. Is any such unity predicable of their forms? Let us seek, in easily-verified facts, for a reply to this question. If a drop of blood be drawn by pricking one's finger and viewed with proper precautions, and under a sufficiently high microscopic power, there will be seen among the innumerable multitude of little circular discoidal bodies or corpuscles which float in it and give it its colour, a comparatively small number of colourless corpuscles of somewhat larger size and very irregular shape. If the drop of blood be kept at the temperature of the body these colourless corpuscles will be seen to exhibit a marvellous activity, changing their

---

* *My* contention is, that the SHORTER *the run* the better for the animal.

forms with great rapidity, drawing in and thrusting out prolongations of their substance, and creeping about as if they were independent organisms.

"The substance which is thus active is a mass of protoplasm, and its activity differs in detail rather than principle from the protoplasm of the nettle. Under sundry circumstances the corpuscle dies and becomes distended into a round mass, in the midst of which is seen a smaller spherical body, which existed but was more or less hidden in the living corpuscle, and is called its *nucleus*. Corpuscles of essentially similar structure are to be found in the skin, in the lining of the mouth, and scattered through the whole framework of the body. Nay, more; in the earliest condition of the human organism, in that state in which it has but just become distinguishable from the egg in which it arises, it is nothing but an aggregation of such corpuscles, and every organ of the body was once no more than such an aggregation.

"Thus a nucleated mass of protoplasm turns out to be what may be termed the structural unit of the human body. As a matter of fact, the body, in its earliest state, is a mere multiple of such units variously modified."

Here, in all its fulness, I have given the Huxleian rendering of his own theory. The noxious leucocyte, the amœba of the blood, is taken as the very creator of man. Here is the first "consequence" (as Herschel would say) of this wondrous theory. Here, is its *very particular application* to man. The physical basis of Death is taken for the physical basis of Life, aye, and *must* be so taken as a logical necessity, unless we refuse to worship the Idol which Prof. Huxley has set upon its legs. For, observe, it was a not unnatural and practically harmless blunder for early physiologists to mistake the amœba of the blood for a valuable constituent of that fluid, but this blunder becomes a fearful power for mischief when it is worked up into a system.

For all who have learned what the leucocyte really is, there could not be a more vivid *reductio ad absurdum*. The protoplasm-theory, when particularized and applied, breaks down completely.

But suppose we *regain* the wild freshness of our morning faith in Professor Huxley and the white corpuscle, and suppose we retain all our confidence that Nature herself teaches thus, and that Huxley truly represents the facts of the case. Let us firmly believe that the white corpuscle is the structural unit of the human body, and a valuable portion of that protoplasm which is the physical Basis of Life, and accompany the Professor into some of the consequences into which this will lead us. Let us hear what appearance Nature presents viewed under this theory. I give his own description:

> " Under these circumstances it may well be asked, How is one mass of nucleated protoplasm to be distinguished from another? Why call one plant and the other animal?
>
> "The only answer is, that so far as form is concerned, plants and animals are not separable,

and that in many cases it is a mere matter of convention, whether we call a given organism an animal or a plant" (p. 128).

He then cites the case of a living body common in one form in tan-pits, and called *Æthalium septicum*, and which, by a kind of biological allotropism (possibly analogous to chemical allotropism or power of appearing in two or more forms, while, like Proteus, retaining identity), is sometimes like a plant and sometimes like an animal in its way of feeding, or "mode of assimilation."

By a curious inversion of the rule of proceeding from the well-known to less known he argues from the imperfectly and *badly* known, to the utter overthrow (philosophically) of one of Nature's most important distinctions.

"Is this a plant, or is it an animal? Is it both or is it neither? Some decide in favour of the last supposition, and establish an intermediate kingdom, a sort of biological No Man's Land for all these questionable forms. But as it is admittedly impossible to draw any distinct boundary line between this No Man's Land and the vegetable world on the one hand, or the animal on the other, it appears to me that this proceeding merely doubles the difficulty, which before was single."*

Here then is the argument; because, forsooth, the limitations of our senses, and the imperfections of our means of research, make us to fail to distinguish always clearly and accurately the boundary line, we must simplify matters by denying that there is any

* For a popular description of this "No Man's Land," see the Introduction to Trouessart's "*Microbes*," &c.

natural boundary at all between the animal and vegetable kingdoms, they are separated only by artifice. But I must give the very words, or the reader will think I am imposing on him.

"Protoplasm, simple or nucleated, is the formal basis of all life. It is the clay of the potter, which, bake it and paint it as he will, remains clay separated by artifice and not by nature, from the commonest brick or sun-dried clod" (p. 129).

Animal and vegetal seem separated only by artifice and not by nature, when that mischievous Puck called the protoplasm-hypothesis squeezes the juice of his "Love in Idleness" into the eyes of the Leaders of physiological thought and makes *natural* confusion worse confounded.

In immediate connection we have a very remarkable instance of the mental confusion, which this theory engenders. I shall call it the "Calc-spar fallacy," and first explain what I mean by this expression, its suitability may then be left to the reader.

If you take a crystal (rhombohedron) of calc-spar, and look through it a mark such as "/" you will see it as "//." Of course it exhibits the phenomenon of double refraction! Gentle reader, pity my simplicity, and let me finish my very simple little illustration in peace. I am not going to bore you with mathematics, at least not beyond simple or compound addition in arithmetic.

Have you ever reflected that there may be as Macbeth does not say—"a calc-spar of the mind?" I mean that the very ablest and acutest minds may exhibit the phenomenon of double refraction. I am not speaking disrespectfully. I love the calc-spar. Cut into the deft shapes of the Nicol's prisms, it forms the ornament of my dearly-loved microscope, the polariscope, and that is by virtue of this very property of double refraction. I tried a very simple and foolish experiment, you can do likewise.

Instead of the single line, which will appear double in the soberest of eyes—take two figures writing them very close together, thus: 12. Now look at them through your spar, and by turning it round you can have all the following varieties of configuration, 2211 , 21 , 21 , 21, 2211, and so on round the
 21 21 21
whole circle. You see how your spar confounds the decimal system. How, if applied to the addition of money, it would make your pence to shillings, without any regard to the values of the different vertical columns—viz., by " direct application" of one figure-value to the other, without remembering or perceiving the necessity of " carrying on" from one to the other, according to the relative values.

Instead of naming our columns units, tens, &c., or pence and shillings, &c., let us call the first column— say "mechanical," the second " chemical," the third " biological," and so on.

I can carry on my result, say from the chemical into the biological column, and so long as I remember the distinction, my work may be correct. But when I use the mental calc-spar, I directly apply the results of chemistry to biology, without noting this distinction.

Let $(C) = 2$ in chemical column, and $(B) = 2$ in biological, let us assume that these two are related in decimal proportion, as the vertical columns in common arithmetic; then our expression $(B) + (C) = 22$, but if you are mentally calcareous, and a good controversialist, and fond of sparring—without altering a single fact of the external world or figure on your paper, you may behold your first column thus : $\{^{B}_{C}\}$; or, instead of 22, you see $\frac{2}{2}$ which make four, and expatiate eloquently on the unscientific, unphilosophical, theological, metaphysical, or what not, minds who cannot see that two and two make four.

Now the reader will know what I mean by the calc-spar, or double refraction fallacy, and we return joyfully to the "Lay Sermon."

> "The statement that a crystal of calc-spar consists of carbonate of lime is quite true, if we only mean that by appropriate processes, it may be resolved into carbonic acid and quicklime. If you pass the same carbonic acid over the very quicklime thus obtained, you will obtain carbonate of lime again ; but it will not be calc-spar, nor anything like it. Can it, therefore, be said that chemical analysis teaches nothing about the chemical composition of calc-spar? Such a statement would be absurd ; but it is hardly more so than the talk one occasionally hears about the uselessness of applying the results of chemical analysis to the living bodies which have yielded them."

Every chemist and biologist must agree to all this. We know there is at least the beginning of a science of animal chemistry which has already given us results of nearly priceless value. But let us carefully note Professor Huxley's method of dealing with the results of chemistry and applying them to biology. We shall find it exhibits that want of perception of breaks, or barriers, or differing values of columns only too familiar to those who have to teach young children their early lessons in arithmetic, and which we have dignified by the name " calc-spar fallacy."

Here are his *ipsissima verba:*

> "Carbon, hydrogen, oxygen, and nitrogen are all lifeless bodies. Of these, carbon and oxygen unite, in certain proportions and under certain conditions, to give rise to carbonic acid; hydrogen and oxygen produce water; nitrogen and hydrogen give rise to ammonia.
>
> "These new compounds, like the elementary bodies of which they are composed, are lifeless. But when they are brought together under certain conditions they give rise to the still more complex body, protoplasm, and this protoplasm exhibits the phenomena of life."

To have been quite fair the Professor should have said that this protoplasm never exhibits a single phenomenon of life unless it is actually in a living organism, or directly taken therefrom. But the calc-spar has evidently been too strong upon him, for he proceeds:

> "I see no break in this series of steps in molecular complication, and I am unable to understand why the language which is applicable to any one term of the series may not be used to any of the others. We think fit to call different kinds of matter carbon, oxygen, hydrogen, and nitrogen, and to speak of the

various powers and activities of these substances, as the properties of the matter of which they are composed.

"If scientific language is to possess a definite and constant signification whenever it is employed, it seems to me that we are logically bound to apply to the protoplasm, or physical basis of life, the same conceptions as those which are held to be legitimate elsewhere [*i.e.*, a 2 is a 2, no matter whether it is in one place or another, as young Hopeful has often blubbered out over his sums]. If the phenomena exhibited by water are its properties, so are those presented by protoplasm, living or dead, its properties.

"If the properties of water may be properly said to result from the nature and disposition of its component molecules, I can find no intelligible ground for refusing to say that the properties of protoplasm result from the nature and disposition of its molecules.

"But I bid you beware that, in accepting these conclusions, you are placing your feet on the first rung of a ladder, which, in most people's estimation, is the reverse of Jacob's, and leads to the antipodes of heaven. It may seem a small thing to admit that the dull vital actions of a fungus or a foraminifer are the properties of their protoplasm, and are the direct results of the nature of the matter of which they are composed.

"But if, as I have endeavoured to prove to

you" [observe the word "prove." If a thing is proved it must remain unalterably true. It cannot change except by its being proved the premisses are false, which, in this case, means the abandonment of the protoplasm theory itself],* " their protoplasm is essentially identical with and most readily converted into that of any animal, I can discover no logical halting place between the admission that such is the case and the further concession that all vital action may, with equal propriety, be said to be the result of the molecular forces of the protoplasm which displays it; and if so, it must be true, in the same sense and to the same extent, that the thoughts to which I am now giving utterance, and your thoughts regarding them, are the expressions of molecular changes in that matter of life which is the source of our other vital phenomena."

The reader will observe that I trouble myself and him with these quotations only so far as they contain " experimental reasoning concerning matter of fact and existence."† I think it can be shown that the experimental reasoning is fallacious, the "facts" relied on are not facts at all, but gross misrepresentations of the phenomena of Nature.‡

* I add this to meet a possible objection to my going back to the "Lay Sermons" (published in 1870). I do so simply because they contain the best description of the protoplasm theory and its logical consequences that I know in the English language.

† Hume, quoted by Huxley.

‡ That the greatest mental acuteness and scientific training are sometimes rather apt to betray their possessors, in simple

One more comparison of the protoplasmic theory with the phenomena of Nature and we have finished this part of the process we have undertaken.

How does this theory bear upon the question of food, especially in regard to human aliment?

Again, in Prof. Huxley's own words, we have the interpretation of the dark sayings of the protoplasmic Sphinx. Whose fault is it that we are so startlingly reminded of the old classical myth of Oidipous, who, by interpreting the Sphinx's famous riddle about human life, not only caused the welcome destruction of the monster who propounded it, but also gained the dreadful privilege of marrying, or rather outraging, his own mother, and remorsefully putting out his own eyes, was self-condemned to blindness.*

There is indeed a sad proof of that blindness to even the common facts of daily life, which comes on those who habitually trifle with the sanctities of Mother Nature, in these truly marvellous words ("*Lay Sermons,*" p. 133).

---

matters of fact, is well illustrated by King Charles II.'s famous inquiry of the Royal Society.

"He asked the cause why a dead fish does not (though a live fish does) add to the weight of a vessel of water. This implies two questions, the first of which many of the philosophers for a time overlooked—viz., 1. Is it a fact? 2. If it be a fact, what can cause it?" (Whateley's "*Logic,*" p. 120.)

None but philosophers would have been taken in by the joke of the Merry Monarch, a courtier would have joined in the joke, a fishwife would have given a rough and ready answer. The king is dead and gone, but the Royal Society is there still, and has many philosophers, as of old.

* See in contrast a charming little work called "*Michael Faraday,*" by T. H. Gladstone, Ph.D., F.R.S. (p. 65), on the

"Hence it appears to be a matter of no great moment what animal or what plant I lay under contribution for protoplasm, and the fact speaks volumes for the general identity of that substance in all living beings.

"I share this catholicity of assimilation with other animals, all of which, so far as we know, could thrive equally well on the protoplasm of any of their fellows, or of any plant, but here" [mark the caution of the philosopher, his language must not be taken as the hyperbole of the mere popular lecturer, he has his sober eye on the limits of his observation; even his omnivorous empire is not *sine fine*] "the assimilative powers of the animal would cease."

"A solution of smelling-salts in water, with an infinitesimal proportion of some other saline matter, contains all the elementary bodies which enter into the composition of protoplasm; but I need hardly say a hogshead of that fluid would not keep a hungry man from starving, nor would it save any animal from a like fate."

Now, I ask any one whether they will try the simple experiment of preaching that part of the "lay sermon" to their cooks, and see it *rationally* carried out. It appears to be a matter of no great moment what animal or what plant I lay under contribution for my dinner—no, don't take me up so literally—for the physical basis of my life.

reverential attitude of that great man of science. "Throughout his life, Michael Faraday appeared as though standing in a reverential attitude towards Nature, Man, and God. Towards Nature, for he regarded the Universe as a vast congeries of facts which would not bend to human theories."

But, jesting apart, what is Nature's clear answer to all this?

It *does matter exceedingly* what plant or what animal you lay under contribution. Nay, if you confine yourself to the best animal food—beef, mutton, &c., unmixed with fruit and vegetables—the loathsome and deadly disease of SCURVY is Nature's reply to the rash theorist, who acts upon the principles of the now popular school of physiology.

To sum up:

1. We have tried the protoplasm theory on the vegetable and animal kingdoms, and it confounds the distinctions Nature is most heedful to set up.

2. We have tried it on man's body, and it forces us to put death for life, and, by a parody of Christian teaching, to love our enemy, under the delusion that he is our friend, and even our creator.

3. We have seen that it confounds the deep distinction laid in a right interpretation of Nature's teaching, between *chemical* and *biological* science.

4. In regard to man's food, it is utterly misleading, and if in any sense a light, it is an *ignis fatuus*, a "light that leads astray."

The result is that we must *modify the statement* to make it correspond to the phenomena of Nature.

Let us see *where* the error lies.

If protoplasm were only a name for a proteid or albuminous substance, containing the elements, carbon, hydrogen, nitrogen, oxygen, sulphur, and phosphorus, in more or less constant proportions, there could be no objection to that name for a compound which undoubtedly exists, and is widely diffused in organisms.

It is indeed evident that the Law of Interchange demands a similar ultimate analysis for the material of which plants and animals are composed. As we have seen in Chapter I., that portion of the mineral world which is in such intimate relation to the vegetable and animal kingdoms, as to form the *immediate environment* of organisms of either kind, must also have a similar ultimate composition. Two States, suppose, which mutually and entirely depend upon each other, existing solely upon the exchange, or rather interchange of commodities, and (suppose) at an equilibrium of profit and loss. If we take the imports and exports in each case and sum them up all together, the results would be identical, though the articles included in exports in one case would be imports in the other, and *vice versâ*. This very imperfectly represents the complete mutual inter-dependence of the three kingdoms of material Nature.

But the term "Protoplasm" ($\pi\rho\hat{\omega}\tau o \varsigma$ first, $\pi\lambda\acute{a}\sigma\sigma\omega$, I form)* is clearly intended to imply that not only the

---

* I give the meaning in Quain's "Anatomy" (Index). Huxley's "Elem. Physiol.," gives "$\pi\lambda\acute{a}\sigma\mu a$," "workmanship." Some critic, whose mind has not travelled beyond his lexicon, may object that this ignores the distinction between "$\pi\lambda\acute{a}\sigma\tau\eta\varsigma$," "the former," and "$\pi\lambda\acute{a}\sigma\mu a$," "the thing formed." Perhaps so. As a matter of fact the word is used in the sense I have described, as I could easily and abundantly prove. If I were writing to be misunderstood by philosophers instead of trying to make a difficult subject as clear as I can to people of plain common sense, I might use, instead of "forces," some term like "potencies," or put the whole expression thus—"Materials supplying also arrangements suited to the manifestation of forces."

The reader ought to be warned that the word "protoplasm" is used in two very different senses. The following quotation

materials but also the *forces* of organic life are contained in the substance so named.

To make it square with the facts, we must express it thus. (I admit rather cumbrously, but it is better to be awkwardly moving on the right path than, with all scientific airs and graces, to be going wrong.)

(*a*) Vegetable substances not deprived of the solar force, "locked up in their compounds," constitute the animal protoplasm—*i.e.*, the material and forces required for animal organisms.

(*b*) Certain inorganic compounds, plus an unknown amount of sun-power, constitutes the vegetable protoplasm—*i.e.*, the material and forces required for vegetable organisms.*

---

from the latest edition (Part I., 1888) of Prof. Foster's "*Text Book of Physiology*" will make these two senses as clear as such needless confusion can be made.

"Protoplasm," in fact, as in the sense in which we are now using it, and shall continue to use it, is a *morphological* term; but it must be borne in mind that the same word protoplasm is also frequently used to denote what we have just now called "the real living substance." The word then embodies a *physiological* idea; so used it may be applied to the living substance of all structures, whatever the microscopical features of those structures; in this sense it cannot at present, and possibly never will be, recognised by the microscope, and our knowledge of its nature must be based on inferences" (p. 5).

* As we have certainly no proof that sun-power only, and not also star, planet (including earth) and moon power, may be concerned in the vital forces of both plants and animals, it may be better to borrow a term from astrology and speak of the "*circumambient*," meaning the whole environment in the fullest sense, which evidently includes the whole material creation. Herbert Spencer says ("*Biology*," p. 85), "literally, the environment means all surrounding space, with the co-existences and sequences contained in it . . . ." Compare also the table at page

Thus we keep before our minds the all-important natural distinctions—(1) between organisms and non-organized matter, and (2) between the mineral, the vegetal, and the animal kingdoms.

Before concluding this chapter, I wish to express my cordial agreement with the philosophic views of Professor Huxley, not that any words of mine could add to his now "old and just renown," but partly as an act of justice, and chiefly because it will bring out more clearly the *gist* of my contention.

To represent his views as materialistic seems to be such an outrage, not only on common honesty, but even on common sense, as to make such objections not worth notice.

I go further and declare my conviction that many of our religious and spiritual teachers are very materialistic compared to him.

I have studied the Bible, with the aid of all the more accessible commentators, I have read my "Plato," under Taylor's neo-platonic guidance, I have, at least, tried to master Swedenborg's teaching, and have mystified myself, I hope usefully, in Jacob Boehme's works, and nowhere do I find a *less materialistic* philosophy than that of Professor Huxley.

I, for my part, know no fuller and better exposition of that saying which expresses the very key-note of Christian doctrine—Καὶ ὁ Λόγος σὰρξ ἐγίνετο*—than in

467 of the same book, where astronomic and geologic changes are reckoned amongst the factors which co-operate in the evolution of life.

* St. John i. 14, and vi. 63. Thought expresses itself in a material form, which latter, in itself, counts for nothing (ἡ σὰρξ οὐκ ὠφελεῖ οὐδέν), though all important as a symbol, making truth "more or less accessible to us."

these words from that "Lay Sermon" to which I have made so many references.

"In itself it is of little moment, whether we express the phenomena of matter in terms of spirit, or the phenomena of spirit in terms of matter: matter may be regarded as a form of thought, thought may be regarded as a property of matter,* each statement has a certain relative truth.

"But with a view to the progress of science the materialistic terminology is in every way to be preferred.

"For it connects thought with the other phenomena of the universe,† and suggests inquiry into the nature of those physical conditions, or concomitants of thought, which are more or less accessible to us, and a knowledge of which may, in future, help us to exercise the same kind of control over the world of thought, as we already possess in respect of the material world; whereas the alternative or spiritualistic terminology is utterly barren, and leads to nothing but obscurity and confusion of ideas.

"Thus there can be little doubt that the further science advances, the more extensively and consistently will all the phenomena of Nature be represented by materialistic formulæ and symbols.

"But the man of science, who, forgetting the

---

* This shortly expresses the distinction between the teaching of Plato and that of St. John the Divine.

† Compare 1 Ep., St. John i., 1st and 2nd verses.

limits of philosophical enquiry, slides from these formulæ and symbols into what is commonly understood by materialism, seems to me to place himself on a level with the mathematician who should mistake the x's and y's with which he works his problems, for real entities—and with this further disadvantage, as compared with the mathematician, that the blunders of the latter are of no practical consequence, while the errors of systematic materialism may paralyse the energies and destroy the beauty of a life."

My objection to the protoplasm theory is that, as usually understood, it is an over-hasty generalization, which has done harm by making us practically ignore the Law of Interchange and its corollaries. It has *explained away* the distinction between the animal and the vegetable.

Science, while it explains away and obliterates merely artificial distinctions, emphasizes, and in explaining increases, natural distinctions, and in spite of all exceptions, such as flesh-eating plants and the numerous tribes of carnivorous animals, that is a true Law of Nature, and good for man to follow, which makes the plant our immediate food provider.

APPENDIX TO CHAPTER VI.

REMARKS ON THE GENERAL REASONING ON WHICH THE CONCLUSIONS OF THE SPECIFIC MICROBE THEORISTS ARE BASED, AND ON SOME OF THE METHODS EMPLOYED BY THEM FOR CLASSIFYING AND IDENTIFYING MICRO-ORGANISMS.

In the Introduction to *Micro-organisms and Disease* Dr. Klein lays down four conditions (taken from Koch's *Die Milzbrand-impfung*) which must be complied with before it "can be said to be satisfactorily proved that a particular infectious disease is due to a particular micro-organism." These are given fully on pages two and three of that work. I here give them briefly in my own words.

The four conditions are that the particular micro-organisms must be

1. Present in the blood or tissues of the man or animal suffering or dead from the disease.

2. Cultivated in suitable media outside the body, and so as to be secure from all possible introduction of other micro-organisms during the process of culture.

3. After having been thus cultivated for several successive generations, they must be introduced into the body of a healthy animal susceptible to the disease, and it must be shown that the animal becomes affected with the particular disease.

4. And, finally, that in this "so affected new

animal the same micro-organisms should again be found."

"A particular micro-organism may probably be the cause of a particular disease, but that really and unmistakably it is so, can only be inferred with certainty when every one of these desiderata have been satisfied" (p. 3).

Now first notice condition 4, which I give fully in Klein's own words. And one question must arise. How can you prove that the so affected new animal was affected solely by the micro-organisms introduced, and not by the dangerous assault on the life, which, as Klein's own book has proved, the *very operation of inoculation in itself is.* Again, though it may be possible to sterilize the artificial media of cultivation, how can the body, blood, and tissues of a living animal be effectually sterilized?

But let us suppose that all these difficulties are successfully met, and that every possible source of error within the limits of these four conditions has been completely excluded—say by repetition of experiment, so that, by the doctrine of chances, the origin of the disease from that particular pathogene, and nothing else, becomes a moral certainty; I say, granting all this, there is a FIFTH CONDITION, unmentioned by Messrs. Klein and Koch, which must be complied with before the experiments come within measurable distance of a satisfactory proof.

It is quite true, as Klein shows, that these four conditions rigidly enforced would simply decimate the most popularly celebrated pathogenetic "proofs," including the evidence for Pasteur's hydrophobia microbe and Koch's own comma-baccillus of cholera,

"*Apparent rari nantes in gurgite vasto.*" But the inexorable logic of fact and natural law disposes even of these hardy and vigorous few that can keep above water when the four conditions are enforced.

This fifth condition is,

That the signs and symptoms of the particular disease should be definite, and such main symptoms so constant that the diagnosis should always be perfectly reliable; nay, it is demanded *for perfect proof* that the disease should be in itself a sort of definite entity, recognizable and definable, otherwise than by its (supposed) connection with the microbe.

The whole reasoning is simply an illustration of the fallacy called "arguing in a circle," for the limit which defines what symptoms shall or shall not be regarded as diagnostic of the particular disease *is this very connection with the microbe*, which has to be proved pathogenetic.

It would force me to bring in medical details, unsuited to this little work, if I were to furnish proofs of this latter statement here, but any student can verify for himself the truth of my remark, and even the "general reader," if he has ever glanced at medical literature, must be aware that the fifth condition never has and never can be complied with.

We next turn to the micro-organisms themselves. How are they identified? We know what a difficult matter is this question of identity (I do not refer to metaphysical difficulties connected with personal identity), but physical identity is a difficult and perplexed subject. We have known the greatest experts have sometimes been deceived, and cases of mistaken identity are not unfrequent. It is true in the case of the microbes there is no attempt to identifying the *individual*, only the species. The "same micro-organism" means the same class, sort, or species of micro-organism.

What is the chief method employed by microbe theorists for identifying their breeds? They identify them much as sheep are distinguished, viz., by being stained by colours, which are understood by their keepers, only with this tremendous difference, that the micro-organisms are chiefly distinguished by the different way in which they submit to the process of being stained, with tar or aniline dyes.

At page 6 of *Micro-organisms and Disease*, we read:

> "Micro-organisms have a great affinity for certain dyes, and therefore these are used with great success to demonstrate their presence, and to differentiate in many instances morphological details which, in the unstained condition, are not discernible."

A most perilous proceeding, full of pitfalls and possible errors, when used for the first investigation of new truths, although every microscopist admits the beauty and value of prepared and stained specimens, for demonstrating facts which *have been already learned*, by a more simple and direct mode of interrogating Nature.

Let us view this process through a mental microscope. Take it as being applied to larger organisms. Remember we must *assume* that these micro-organisms *are living creatures*, they would not be organisms otherwise. Remember also that, though from a proto-plasmic point of view "it is a mere matter of convention, whether we call a given organism an animal or a plant" (Huxley), they must come under a category which *includes* all plants and animals. Perhaps some may be plants and some animals, and a field in which these organisms are artifically collected, fed, and cultivated may, in an *enlarged* sense, be aptly compared either to Kew Gardens or the Zoo.

Who but a madman would go if he could escape the keepers, and dabble these organisms with tar dyes, and mark their behaviour under them, as if that afforded a ground for scientific identification and classification?

Does the folly diminish in direct or in inverse relation to the size of the organism, or does it not rather remain constant, and is it not utterly absurd to attempt to classify organisms biologically (not chemically, mark you, but biologically—*i.e.*, *as living creatures*), by their behaviour under chemical reagents?

By chemical processes, whether employed on organic or inorganic materials, we get *chemical results* (and valuable and precious they may be in their own line); but biological results must be obtained by biological means;—by means of *natural selections* and *affinities*, and cultivations and breedings—*good* breedings (in every sense), not forcing but obeying the rules of life; not outraging but wooing the modesty of Nature.

Then again look at the method of naming which these experimenters adopt. We can more or less accurately gauge the condition of any science by its nomenclature.

Their method has a curious resemblance to Hamlet's

description to Ophelia of the method of the "Prologue" players to the murder of Gonzago. After the dumb show of *injecting poison*, in reply to Ophelia's questions, Hamlet says:

"*Ham.* We shall know by this fellow. These players cannot keep counsel; they tell all."

"*Oph.* Will he tell us what this show meant?"

"*Ham.* Ay, or any show that you'll show him. Be not you ashamed to show. He'll not shame to tell you what it means."

Take one example. Nature (as if forced to act in accordance to Hamlet's mocking advice) *showed* a tiny point—an organism "of about the .0025 part of a milimetre in diameter, spherical or oval, and of a bright red colour" (*Klein* p. 63).

No less than two of the "players" are down upon that show, and, though they could not tell the meaning (for Nature refuses to such men and such methods her meanings), they can *name*.

Cohn calls it, "Clathrocystis roseo-persicina," and Lankester comes in with a second name—nay, I must be careful. It appears to be a matter of the utmost moment who *calls out first* in this cruel game, and I really cannot say whether Cohn or Lankester saw this particular show first, and we may believe that both may divide the crown.

The alternative name is Bacterium *rubescens*. How suggestive is this inceptive *rubescens*; the bacterium, that is "*beginning to blush.*"

Outraged humanity may cause that blush to extend

from that tiny point until universal scorn and shame will put a stop to these abominable absurdities. Say it again, Clathrocystis roseo-persicina. Does it trouble your tongue? Think what tortures that chamber-full of poor dogs and rabbits and guinea-pigs, and countless other patients in similar chambers, must have throbbed through, while that useful name was being piled together, and that "show" elaborated.*

"Science is, I believe, nothing but *trained and organized common sense*," says Professor Huxley. Be it so.

Then in what department of science shall we place these methods and these names?

Whence comes this method of testing organisms by chemical re-agents (not chemically analysing, according to the fair methods of animal chemistry), but classifying them *as organisms* by their behaviour under aniline dyes!

Read again this description of protoplasm, and you see from what these players take their cue.

> "Protoplasm, simple or nucleated, is the formal basis of all life. It is the clay of the potter, which, bake it and paint it as he will, remains clay separated by artifice and not by Nature from the commonest brick and sun-dried clod" (Huxley's *Lay Sermons*).

---

* The spirit of Domine Sampson, if one could so wrong that kindly pedant as to give him a part in this cruel fellowship, appears to have stood godfather in one instance, closely allied to the bacterium rubescens, as being *chromogenic*, this is the micrococcus prodigiosus. It is blood-red, and "the cells are the smallest of all pigment micrococi."

"He grinned like an ogre, swung his arms like the sails of a windmill, shouted ' Prodigious' till the roof rung to his raptures." "*Guy Mannering*," chap. xviii.

Thus, under the protoplasm hypothesis, it becomes quite plain why we have this medley of chemical and biological methods. This wild phantasmagoria of microscopic forms, truly " separated by artifice and not by Nature," and " like the painted clay," according to the colours they assume under aniline dyes, is simply an exhalation arising out of that Serbonian bog which science has left too long undrained on the confines of her cultivated territories, where chemistry loses the exactness of its rational formulæ,* and biology is sunk in utter confusion of that most important boundary line, which separates between the living and the dead.

* Rational formulæ " are intended to indicate the chemical nature of the compound and to express the relations in which it stands to other bodies." (Roscoe's and Schorlemmer's *Treatise on Chemistry.* Vol. III., Part I., p. 112.) It is evident that protoplasm, being indefinite in its composition, *can* have no rational formula. Its empirical formula is variable and inconstant.

CHAPTER VII.

HOW THE THREE DESIDERATA OF THE CELEBRATED PHYSICIAN SYDENHAM HAVE BEEN DISCOVERED AFTER TWO HUNDRED YEARS OF WAITING.

THE illustrious Sydenham, after whom some of the most learned of our medical societies delight to name themselves,* has somewhere spoken of three things which were still wanting in his day, and much to be desired. The third of these especially he would, he declared, hail with enthusiasm.†

Two hundred years have rolled away since the death of that great physician, and Sydenham's desiderata are practically things still wanting in all medical teaching. If the spirit of that great man now survives amongst medical men, how gladly would they like him bring eager assent and joyous acceptance to greet one, through whom these long-sought and most desirable things have at last been discovered, and offered to all who will avail themselves of the benefit.

These desiderata are :—

1. An account of diseases, true in description, and

* The Sydenham Society, instituted 1843, and the New Syd. Society, 1858.

† His words are: "Jam vero si quærat aliquis an ad praedicta in Arte Medica desiderata duo (veram scilicet et genuinam morborum Historiam, et certam confirmatamque Medendi Methodum) non etiam accedat tertium illud, remediorum nempe Specificorum inventio; assentientem me habet et in vota festinantem (Praef. Ed. Tert., *Observ. Med. Syd. Om. Oper.*, p. 18).

correctly tracing their genesis or origin ("*veram et genuinam morborum Historiam*").

2. A reliable and proven method of healing ("*certam confirmatamque Medendi Methodum*").

3. The discovery of specific remedies ("*remediorum nempe specificorum inventio.*")

It is my pleasing task to give an outline (though only incidentally, as so far as they belong to my subject, which is purely biological) of these discoveries.

For a full account of the system, which is at once a simple and easy *Practice*, and also a THEORY in the full and grand original sense of that word θεωρία, a BEHOLDING of the very truth and method of Nature,* I must refer the reader to the discoverer himself.

For all in this chapter, and indeed in this little book, that is not marked fully *as quotation*, with references given to the source from which it is extracted, I, the writer, am alone responsible. Even the headings and arrangement of the extracts are entirely my own, so I must in fairness refer the reader to the work from which these extracts are taken, and warn him that, although I have myself *independently verified every statement* contained in these extracts, and seen with my own eyes all herein described as seen, and proved by my own experiment, or in my own experience, every fact herein described as fact, it is only a small

---

* Such as Sydenham attributes to Hippocrates: "Atque in his fere stetit magna illa Divini Senis θεωρία, non ab irrito lascivientis phantasiæ conamine desumpta, ceu vana ægrorum insomnia, sed legitimam exhibens historiam earum Naturæ operationum quas in hominum morbis.edit. Cum vero dicta θεωρία nihil esset aliud quam EXQUISITA NATURÆ DESCRIPTIO . . . ." (*Op. Om.*, p. 14.)

sample extracted for a special purpose, and not pretending to be even an outline sketch of the "THEORY," and *giving nothing whatever of the practical portion of the system* that is here presented for his consideration.

Before beginning a series of quotations from his writings, a few words about the scientific life and training of Joseph Wallace may help to explain something both of the nature of his discoveries and how he was led to them.

In his early youth he was much given to experimenting—electricity and chemistry being his favourite sciences. He was, as a very young man, engaged in the business of malting and distilling. Thus he was led to study the theory and practice of fermentation. Having been early initiated into all the secrets of wort-making and brewing, he learned the exact limits of temperature at which the yeast displays its varied activities.

His special training, aided by luminous commonsense—the best of all scientific endowments—saved him from falling into an error, which, like the Teutoburgian passes to the gallant legions of the Emperor Augustus, under the brave but unfortunate Quintilius Varus, has proved a hopeless stumbling-block to hosts of physiologists, and some of their most illustrious leaders.

Every reader of works on physiology, especially by the popular authorities, can see for himself a very ludicrous thing. It may not strike the reader, it does not appear to have been noticed by thousands of teachers, but once pointed out, every intelligent reader wonders why he has never detected the absurdity for himself.

Note how often observations are made at the temperature of about 60° Fah. on organisms or organic fluids whose normal living temperature is about 98·6° Fah. (blood heat).

What would be thought of a gardener who essayed the cultivation and study of tropical orchids at a temperature of *zero?* Yet the absurdity would be much less; for orchids flourish in a wide range of temperature, and no great biological theory affecting the welfare of humanity is rested upon their habits, not even in the creative work of Darwin on *Orchids*.

Now the yeast organism is literally *sensitive to a degree*, even of the Fahrenheit scale. It is simply torpid at 60°, and many degrees higher.

The world will have cause to rejoice that Wallace's early business brought him into an intimate acquaintance with yeast and its strange habits of life.

Later in life Wallace was engaged in another business, which at first sight would seem to be entirely out of the way of all physiological discovery, and yet in truth was of the utmost service as a training. This was the manufacture of embroidery. His eye received a thorough training through the immense number of specimens of work which passed under his inspection as a master-manufacturer in that industry. Frequent attempts to pass off previously used designs would be made, and could only be guarded against by constant vigilance. Thus not only the power of distinguishing varieties of design, in closely resembling forms, but also the facility of recognising identity and the memory of forms were cultivated to a remarkable degree.

Every student of the history of human knowledge must know that the grandest discoveries in any branch of science are made not by the professed and professionally trained student, still less by the salaried professor, but by those who would be regarded as outsiders.

The history of astronomy tells how man's deepest convictions, supported by the evidence of the senses as well as by the interpretations of the Christian faith itself, were revolutionized by a church dignitary—Canon Copernicus. Or if it does not seem so strange, that the learned leisure of an ecclesiastic should be thus employed, at a period, when simply to be able to read, procured the "benefit of clergy," let us take a smaller instance, but perhaps more remarkable, in the incongruity it presents between position and pursuits. Jeremiah Horrox was a young curate of eighteen,[*] but even at that early age a master of the astronomy of his time. Comparing different tables with his own observations of the planet Venus he found a transit across the sun was to be expected on Dec. 4th, 1639.

> "Unfortunately the day was Sunday, and his clerical duties prevented his seeing the ingress of the planet upon the solar disk, a circumstance science has mourned for a century past, and will have reason to mourn for a century to come."[†]

Or take a modern instance—that of the militia band-conductor and church organist, who became one of the greatest of astronomers, and the founder of a family of discoverers, whose name is not only enrolled in the solar system itself, but by whose laborious

---

[*] I give this on the authority of Professor Newcombe.

[†] *Popular Astronomy*, by Simon Newcombe, Professor U.S. Naval University. Second edit., p. 176.

genius, with its infinite capacity for taking pains, the Titanic task of determining the form and plan of the WHOLE VISIBLE CREATION, including the farthest telescopic star or nebula, has been at least attempted.*

I cite another instance which comes nearer to the science of biology—one whom, in allusion not only to his Christian name, but much more to his character and attainments, we may regard as the *archangel* of physical and chemical science—Michael Faraday.

The blacksmith's forge and the bookbinder's workshop seem at first sight unsuitable schools for the training of the most reverential, careful, and delicate experimenter and observer that ever was permitted to behold and expound the mysteries of Nature.

Now I cite these cases for this purpose—not simply to prove that great discoverers had often what seemed an incongruous training—a paltry truism not worth the mention—but to show that, under this seeming incongruity, there is really the most exquisite grace of congruity.

For no one who has learned in any degree to recognise the order which true science reveals as existing throughout the universe—how under that seeming disorder, "*quem dixere chaos*"—which is not the *chronological*, but rather the *logical* antecedent of the cosmos of beauty and order, can believe that Nature leaves to chance the training of her interpreters. When biography is more of a science, when it becomes more truly *biological*, the life-history of

---

* Sir William Herschel was organist of the Octagon Chapel, at Bath, when he commenced his career of discovery. His sister, Miss Caroline Herschel, assisted, and his son, Sir John Herschel, carried still further the labours and discoveries in astronomy.

every human being will be found as strictly in accordance with laws, *i.e.*, as capable as being grouped and classified and calculated as, let us say, the apparently erratic course of a comet; and the greater and lesser lights which rule the day or night of human discovery in the material and spiritual worlds, perceptible or non-perceptible to merely bodily senses (represented by the dayside and nightside of Nature), will be found at least as reducible to calculation as the courses of the sun and moon are now by the astronomer.

"Unborn the hands but born they are to be," which shall reduce the spiritual universe into as conspicuous order as we can now behold in the material universe, by the labours of Newton and his followers on the same and kindred branches of the great tree of universal science.

Be all this as it may, a connection can be distinctly traced between the training and the discovery. It may seem to the discoverer himself as a thing he can come upon by chance, but the path on which he was treading when he met with that thing—the luminous conception or pregnant Fact which constituted the Discovery—was *not* the result of Chance. We may safely assume that on no other path would he have come upon it, or at all events have seen it in the same relation; the object, if seen at all, would have been seen from a different point of view. In other words, that particular discovery would not have been made.

In the case of Faraday,[*] we can distinctly trace a

---

[*] The writer cannot help expressing his thankfulness to his father for putting into his hands Faraday's *Chemical Manipulation*, telling him he would there learn how to make a most

casual relation between his early training and his subsequent career; so, too, in the case of Joseph Wallace. What training could be better for a microscopist than to have his eye trained to distinguish form, and his mind trained to remember?

The writer can testify, from some little experience of his own, that the forms presented under the microscope are exceedingly bewildering to the untrained eye. What special training have most of our biologists ever received? None whatever, except in the very course of their work, like the training a young clergyman generally receives in preaching, when he has to practise upon the patience of his congregation, so the patients in our hospitals and a miserable multitude of dogs, cats, guinea-pigs and rabbits, fowls and pigeons, have to suffer all the refined cruelties of scientific torture, while the medical biologists are slowly training their eyes at the expense of their hearts, and of every chivalrous sentiment of true manliness and feeling of humanity.* Dr. Becker's field

---

delicate chemical balance out of the cheap materials of a small brass plate, a lath, and a sewing needle. This and many a similar lesson, such as the use of small tubes in chemical operations, where truth and not display is the student's aim, make this work of priceless value to the beginner in chemistry, and are themselves the fruits of Faraday's *training*.

* See Klein's *Micro-organisms and Disease*).

"Dr. Becker then injected a small quantity of the same fluid" (viz., putrid pus matter) "into the jugular veins of fifteen rabbits, after having some days before fractured or bruised the bone of one of the hind legs" (p. 81).

Again, at p. 71, we read in regard to experiments on human creatures:

"Orth cultivated these micrococci (viz., of erysipelas) artificially, and with such cultures produced by inoculation

of fame is the laboratory of the Berlin Imperial Sanitary Office, and there are many other sanitary offices and charitable institutions similarly used for the pursuit of Health, by the propagation of disease.

All the thinking and feeling world must acknowledge that a rational method of healing, and a scientific account of the origin of disease, are eminently desirable, and will agree with the illustrious Sydenham, whose fame is untarnished by any voluntary production of disease, though he cannot be said to be really successful in its cure. He could not cure himself of the gout, from which he suffered for thirty years, of which with some other equally painful complaints he died.*

The modern method is to fight disease, " even as Jannes and Jambres withstood Moses," by imitating and re-producing the acts of the plague-sender.

It is a feeble and shortsighted policy; and in the

---

erysipelas in rabbits. Fehleisen placed this beyond doubt, inasmuch as he produced successive cultures of these micrococci (derived from the lymphatics of erysipelatous human skin), and then by re-inoculation produced the disease not only in rabbits but also in man."

These extracts are interesting, especially when we read Klein's own remarks, given on the same page. There is a charming unity of sentiment amongst these professional brethren, but as opening out new and hitherto undreamed of ways of doing good.

"These inoculations were justifiable because they were undertaken with a view to cure certain tumours. Thus, one case of lupus, one case of cancer, one case of sarcoma, were considerably affected, and to the good of the patient." (*Klein*, p. 71.) We ask, what good?

* " Podagra inde a triginta annis laborabat Sydenhamius (Op. Om., p. 15).

end will, as we trust and believe, aid in the deliverance of a people, and not of one nation only, but in the establishment of a rational as well as truly philanthropic method of healing throughout the world.

The patience and credulity of the world is nearly inexhaustible, but even credulity has its limits, and patience may be at last exhausted.

Men have stood to have their children inoculated with a cultivated small-pox, to procure immunity from natural small-pox, but now, when all supporters of vaccination are bound to own that to be true to the principles of this mighty method of combating disease by propagating it in a milder form, there ought to be a separate vaccination or inoculation for every known or at least prevalent form of disease. Milder dog-madness as a prophylactic against hydrophobia, cultivated cholera to cure cholera, typhoid to cure typhoid, and so on through the category.

The nightmare weight of this result of modern teaching in medicine must rouse the public even out of their drugged and charmed sleep, and once roused to thought, action will be swift, and deliverance will come at last.

I now give an outline of the opposite method, taking extracts which bear upon my subject. Once again I remind the reader that neither in this work of mine, nor in any work which has yet been published, is there anything approaching a satisfactory account of these discoveries and the scientific system of healing founded upon them.

(1.)

DESCRIPTION OF THE WHITE CORPUSCLE.

"The way in which I have observed the white corpuscle to comport itself in the blood is as follows: when one is selected out for observation, it will be seen to writhe and wriggle into various shapes, will become serrated all round, or only at one side, but always in motion; ultimately, it will extend a finger-like prolongation of its substance, and inclosing one of the largest of the nuclei, which may have been seen previously moving about inside the parent cell, and which moving towards the finger point, forms into a small round ball, joined to the main body by a narrow neck, this breaks off completely, or rather is thrown out like an egg from a bird, when the little youngster may be seen sailing off briskly on his own account, and feeding upon the red pabulum (crassamentum) of the blood (which adds the colour to his substance), resting here and there to feed as he goes, as wayward as any other young animal. It can, under favourable circumstances, in a little time be seen *distinctly growing larger* under the eye; but this is not all, for the parent writhes and wriggles again, another is thrown off, and still another, and another may be seen to go off without lessening the bulk of the parent-cell, except for the moment. Each youngster goes on an independent track for itself, and several can still be seen in various stages of development moving about in the parent cell, and as one leaves, another very minute nucleus will be seen just coming into view : thus favouring the presumption that this multiplying process may go on under congenial conditions *ad infinitum*. . . .

"The white corpuscle is to be seen of every size, from the germ just left the parent to the full-grown cell, exceptional ones of which I have seen containing as many as thirty-five germ corpuscles, or nuclei."* (*Physianthrophy*, p. 125.)

(2.)

THE WONDERFUL METAMORPHOSIS REQUIRED BEFORE A LEUCOCYTE COULD CHANGE INTO A RED CORPUSCLE, LONG LOOKED OUT FOR BY HOSTS OF MICROSCOPIC OBSERVERS, GENERALLY ASSUMED BY PROFESSORS OF PHYSIOLOGY, &c., BUT NEVER ONCE SEEN, OR ANY STAGE OF IT.

"Amongst the hosts of microscopists not one so far (although their minds have been *driven* into this groove), has observed a transitory phase of change of that wonderful metamorphosis which *must* of necessity take place before the 'white corpuscle' can be changed into the red.

"For it must first cast off its cell-wall, lessen its size, alter its configuration, internal organisation, and specific gravity. It must change its colour, lose its powers of locomotion, contortion, irritability, food assimilation, growth, and multiplication, or propagation—as also its power to exude through the mucous membranes and blood vessels which the red corpuscle, though smaller, is unable to do without actual rupture of the part—and finally, when paralyzed, to form itself into the *penicillium glaucum* fungus which is seen in great quantity in the expectoration of those dying in consumption as well as in various parts of the body in other diseases" (p. 126).

(3.)

THE LEUCOCYTE AND PENICILLIUM GLAUCUM IDENTIFIED.

"In the sputa of those *recovering* from phthisis, this fungus gradually lessens in quantity, and great masses of white corpuscles come away instead; in this state the *penicillium glaucum* can be clearly seen in process of formation from the white corpuscles, which can be observed arranging themselves in rows just before cohering, and according to size, forming stem or branch; some cohering in part, the rest perfectly organized—up to fructification, in fact.

"PHYSICIAN, HEAL THYSELF."

"Having been consumptive myself for many years until very lately, I examined my own blood very frequently and minutely, and observed that when the disease was in the ascendant the greater the proportion of these white corpuscles to be seen in the blood, the more tenacious was the sputum, or penicillium, and the weaker and worse in health I felt; but when the disease was being overcome, large quantities of the white corpuscles came away so easily that they seemed to have lost their cohesive power, and their relative numbers lessened in the blood in proportion."

ONLY ONE IN FOUR HUNDRED IN PLACE OF ONE IN FORTY LEUCOCYTES TO RED CORPUSCLES.

"At present very few indeed are to be seen; only one in four hundred of the red in place of ONE in FORTY, two or three years ago" [written in 1869]. A note is added—

"A year or two after this, on examining the blood, I found only one white corpuscle in five thousand red corpuscles, when I never had such exuberant health, mentally and physically."

(4.)

IMPROVEMENT IN HEALTH AND INCREASE OF LEUCOCYTES IN INVERSE RATIO.

"The general health has improved in inverse ratio to the number of white corpuscles seen in the blood. The same results I have observed in hundreds of others" (p. 127).

(5.)

THE LEUCOCYTE AND YEAST-ORGANISM IDENTIFIED.

At p. 7 of *Physianthropy*, there is a popularly-expressed account of a ready method of trying experiments which most inadequately represents the real severity of the test conditions under which all experiments in the cultivation of micro-organisms must be conducted.

The result of carefully conducted experiments may be thus given.

In thoroughly sterilized media (such as grape sugar and water, or apple-juice) leucocytes can be directly cultivated out of brewer's yeast, every precaution known to the microbe cultivators' science to prevent contamination or error being rigidly observed.

But a still better mode of experimenting, in my opinion, is the more strictly biological method, viz., watching and noting the phenomena presented spontaneously by the living (human) body, when leucocytes are cultivated in its tissues and blood.

I do not justify any experiments in the production of disease, initiated for the purpose of gaining information. This is a doing of evil that good may come, out of which no true science, no real good, has ever come. But the practice of inoculating with leucocytes is now so extensively practised under the name of vaccination, that an ample field, white already to the harvest of disease-phenomena, is ready for the gathering. Any careful observer can glean useful experience as to the best modes of propagating disease by observing the effects of vaccination.

Vaccine lymph, according as it is more or less pure, is, in like degree, a more or simple cultivation of the leucocyte.

It will not, I hope, be regarded as a wanton experiment in disease-production that once or twice brewer's yeast has been used for vaccination. I have never tried this experiment myself. In my student days I have vaccinated very many hundreds of children, but then I did it, as many others do who follow the ordinary medical teaching and practice, without in the least knowing the horrible nature of these acts. Now nothing would induce me to pollute with the unclean thing the opening promise of an infant's life.

A short quotation gives the result of a sort of experiment which will, I hope, never need to be repeated:

> "The inoculation of brewer's yeast into the blood will produce similar results" [to the inoculation of leucocytes] (see *Phys.*, p. 112).

(6.)

THE LEUCOCYTE AND "PUS."

"Pus, or the matter which we see in boils, ulcers, and abscesses, is nothing but the white corpuscle variously modified by stagnation, position, influence of the temperament of the individual, and the effects of drugs brought to bear upon the organism, whether we see it in an irregular mass, or simple germinal matter, and when boils, eruptions, or ulcers form on the surface, it is Nature's effort* to expel the blood poisons in a mass, and should not be repelled but assisted, otherwise re-inoculation will be the consequence."

TWO PRINCIPAL CLASSES OR KINDS OF PUS MATTER DIVIDED ACCORDING TO ITS ORIGIN AND EFFECTS.

"Pus matter proper is distinctively to be divided into two classes or kinds.

"The first kind is that which is engendered *de novo*, by the stomach receiving into it, for the purpose of digestion, partially decomposed or fermented foods and drinks in which *yeast* forms an *active, living* element. These yeast animalcula (the younger or smaller broods particularly) insinuate themselves readily into the capillaries, and pass into the life current. . . . The yeast lives at the expense of the blood proper, and thus establishes itself in the human organism as the *physical basis of death*. This is the '*white corpuscle*' of physiology.

"The second kind of pus matter is that which has been hereditarily transmitted from diseased parents, or acquired during life from direct inoculation by means of vaccination or other forms of blood-poisoning. The yeast germs are, in these cases, consequently older and more degraded, from their having been

---

* It is interesting to compare with this quite independent observation direct from Nature's teaching the wise anticipation of Sydenham—" nihil aliud quam, Naturæ conamen" (see title-page of this book).

driven in the first place to the surface for expulsion. . . .

"Yeast or pus, as engendered directly from that which is absorbed from the food in digestion, Nature is able as a rule, under ordinary circumstances, to expel daily almost as rapidly as it is produced in the blood. But not so the inoculated pus; nothing short of a radical change in the mode of life, or an epidemic state of the atmosphere, and an eruption of small-pox, will eradicate it, because it is a superadded load of a more degraded form of the parasite" (p. 112).

ORIGIN OF DISEASES AND FOURFOLD CLASSIFICATION OF THEM.

FIRST FORM.—**Simple inflammatory form, a "fermentation" caused by leucocytes.**

"When diseases are traced back to their true origin, it will be found that they have all sprung from one fundamental abnormal condition, which may be designated as the febrile or inflammatory, as in colds, simple fever rheumatic, and other inflammatory fevers, &c. This condition is caused by using animal flesh, fermenting and fermented foods and drinks, in all of which are to be found the white corpuscle or yeast animalcule.

SECOND FORM.—**The same (first form) complicated by drugging, or any hindrance to Nature's efforts at expulsion, thereby compelling a reabsorption.**

"The next stage of disease would be that caused by drugging and other maltreatment of this primary condition, which, instead of aiding Nature to more quickly eradicate or eliminate that ferment, which in this first condition she is always labouring to do, we so encumber her by our ignorant interference, that she not only becomes totally unable to expel the ferment, but is compelled to reabsorb it and the drugs as well." . . .

"Thus is created a chronic or permanently diseased state, in which generally one or more of the vital organs become implicated."

"Nature being thus prevented from throwing out the pent-up pus matter by the skin, in the form of small-pox or other eruption [*pustules*], she is compelled of necessity to segregate it in colonies in various non-sensitive parts of the organism in the form of glandular swellings and tumours, which, in consequence of their being not so fluid as the blood, lessens its (the pus matters) power or self-multiplication, or she may select one or other of the vital organs (and she always selects the strongest) through which to directly eliminate the pent-up matter."

"In this second condition the microscope reveals a change having taken place in the size and behaviour of the white corpuscle. Permanently imprisoned in its abode, so to speak, it necessarily becomes abnormally large and bloated, or more degraded in comparison with the first or acute febrile condition" (p. 147).

THIRD FORM.—**The same (first form) complicated by stimulants of the Fusel oil type.**

"This [stimulation] drives back the moving matter for a time to the transient relief of the patient. . . .

"Stimulation merely adds to the disease, making it more chronic or confirmed at the expense of the vitality and term of life. Stimulants need not necessarily be confined to alcohol, for the worst forms are those bought at the drug-shop, such as opium, morphia, chloral, etc. . . .

"This condition may also be inherited as well as the second condition."

FOURTH FORM.—**The gouty habit.**

"The next condition of any great importance is the gouty habit of body. This is generally inherited. It is usually originated by indulgence in fermented grape juice, or wine," . . . [The process of fermentation] "so disintegrates or decomposes the juice of the grape that it sets free its earthy elements (potash, etc.), from which we get the chemical tartrate of potash (cream of tartar)."

This process is closely followed within the system, when the ferment is present.

"Now potash, it is well known, when taken into the system, seizes upon the animal oil, which forms our reserve force, and turns it into soap. This being dirt or matter out of place, it has, with much waste of the remaining life

> force, to be expelled. When this is impossible, the potash permeates the osseous system, chiefly about the joints, defrauding us of their natural lubricant, and acting as a foreign irritant upon the synovial membrane. . . . Thus gout is created (p. 150)."*

With certain exceptions, such as worms, whooping-cough, cholera, and of course mechanical injuries, "these four conditions cover all the ground of disease incidental to humanity."

> " It is not our fault that these varied expressions of *one definite diseased condition* have had such an infinite variety of names given to them by all previous observers, but it has been our misfortune that none of these have hitherto been able to penetrate the mystery that has for thousands of years been hanging over all things relating to disease and the death principle in man, and so to reduce the ever-increasing chaos into something like law and order (p. 151)."

I make no further comment upon the above extracts. They will speak for themselves. But on the interesting question of the identification of the yeast, the white corpuscle and the blue mould, I wish to present the reader with the best and latest information from recognized and high authorities on the subject.

The identity of the blue mould and all its vinous-fermentation-producing congeners, can be actually

---

* I again remind the reader that all purely medical details are rigidly excluded from this little book. For such and the remedies he must refer to the sources from which these extracts are taken.

deduced out of the observations recorded in the last edition (1889) of *Practical Biology* (Huxley and Martin, extended by Howes and Scott).

I say this advisedly, and in spite of the fact that every resource of language and device of naming has been employed to make artificial distinctions. For instance, the mould that prefers to grow on apricot jam is distinguished by the name Eurotium Aspergillus glaucus, which suggests no affinity to the Penicillium. We are warned (p. 418) that

> "pullulating cells, resembling Torulæ, are not unfrequently derived from the conidia of Pencillium, and many other of the lower fungi, but they must not be confounded with true yeast."

Again, of another mould that sometimes forms on wet and warm bread—and called Mucor Stolonifer, we read that if submerged in a saccharine liquid, it multiplies by budding, after the manner of Torulæ.

> "This '*Mucor-Torula*,' functionally as well as morphologically, bears a resemblance to the yeast plant, from which however its life history shows it to be quite distinct."

In fact the chapter is full of warnings against confounding these moulds with yeast, that I might have feared that after all it was a case of mistaken identity, so clever are the disguises under which yeast appears.

One golden sentence from the same book sets the whole matter at rest, and shows that all these distinctions are truly microscopic, and much less than that, (say) between the different sorts of geraniums. At page 376 we read :

"A saccharine solution will not ferment spontaneously. If it begins to ferment, yeast has undoubtedly got into it in some way or other."

So these mouldly Torulæ, if not yeast themselves, must be *possessed* by *yeast*, because they certainly cause fermentation. "These *Torulæ* are the 'particles' in the yeast which have the power of provoking fermentation in sugar" (p. 379), and the Torulæ of the mould-family do likewise.

But the *Amœba* is an animal, and the *Torula* in two ways proves itself to be a plant. [First by its coat, and secondly, by its power of 'catching a Tartrate' and turning it to its own use.]

"Torula is an indubitable plant for two reasons. In the first place, its protoplasm is invested by a cellulose coat, and thus has the distinctive character of a vegetable cell. Secondly, it possesses the power of constructing Protein out of such a compound as Ammonium Tartrate, and this power of manufacturing Protein is distinctively a vegetable peculiarity. Torula then is a plant, but it contains neither starch nor chlorophyll, and it cannot obtain the whole of its food from inorganic compounds, thus differing widely from the green plants. On the other hand it is in these respects at one with the great group of *Fungi*. Like many of the latter its life is wholly independent of light, and in this respect again it differs from the green plants" (p. 382).

A little further on we read (p. 383):

"It has been further ascertained that Torulæ

flourish remarkably in solutions in which sugar and pepsin replace the Ammonium Tartrate. In this case the nitrogen of their protein compounds must be derived from the pepsin ; and it would seem that the mode of nutrition of such Torulæ approaches that of animals."

Thus we have the newest and highest authority for the statement that the essential identity of the yeast, the blue mould and similar moulds and the leucocyte *cannot be disproved*. I do not claim it as fully admitted, but that the accepted scientific teachers (apart from J. Wallace) are in an *agnostic* state (*i.e.*, neither affirming nor capable of denying) on the subject.

The thoughtful reader will compare these facts (which have been completely verified, and which *prove* the polymorphism of yeast) with the behaviour of the leucocyte in the human body. Outside the body we see the Torula like a plant living on inorganic matter, and so also inside the drugged body, the leucocyte becomes the organizer of drugs into the fell nutriment of disease. It imparts its own noxious vitality to the poisons, and receives in return from them greater permanency, and a deadlier virulence.*

---

* It makes no difference to our argument whether we regard the several fermentative organisms as varieties of one species, or as being of quite distinct species. In the yeast-world, with its threefold and even fourfold *mode*, and its amazing *rate* of reproduction, most of the Darwinian agencies which originate new species, (such as natural or artificial selection, *appetite*-selection, survival of fittest to the environment, &c.) have ample scope for their operation, and that in a few days. We must measure time, as regards development of species, by the succession of generations, or generative phases, in the reproduction of more or less similar individuals, and hence a month or so would fairly correspond to a *geologic period*, in the differentiation into species of the ordinary flora and fauna. This remark also holds good if it be extended from *species* to *genera*.

CHAPTER VIII.

THE BANQUET OF ALMA, OR DIET OF HEALTH, NOT A MEAGRE FARE, BUT WHILE SO CHEAP AS TO MAKE STARVATION ALMOST AN IMPOSSIBILITY, WHEN ONCE THE TRUTH IS FULLY KNOWN, CAN BE EVEN LUXURIOUS.

LET us now bring the facts we have been discussing to bear upon our food.

To recapitulate the foregoing chapters.

Man as an animal comes under the Law of Interchange, which teaches us that the plant is our proper food-preparer, and that the more directly we obey that law the better for us, because, though the *materials* may be forthcoming in flesh and mineral, the forces necessary are either diminished or altogether wanting.

In addition to all this, the microscope reveals to us, as present in the *decaying* vegetable, and in all even the freshly killed and most healthy animal, the amœba of the blood, or leucocyte. Recognizing the extreme value of Nature's "Captain of the Guard," who is also, as in ancient Egypt, the Chief Executioner,[*] viz., the LEUCOCYTE, in his proper place, as head of the great army of Scavengers, and so an all-important Sanitary officer in the commonwealth of Nature, we yet do not desire his services before the time.

Therefore the "honest sonsie face" of Burn's "Great chieftain of the puddin' race," and the perhaps equally honest-looking round of the "roast beef of old

---

[*] Joseph's Potiphar was "Captain of the Guard" to King Pharaoh, the title means strictly, "Chief Executioner."

England," have to be rejected simply on the ground that they are not so honest as they seem. They are like the "honest Iago," and when most trusted are apt to betray the heart that trusts in them. The microscope detects, under their bluff friendliness, the deceitful leer of the leucocyte.

Wide as is my hospitality, at least in intention, I should not care to sit at table with my own destined hangman, "*not* if I *know* it."

But still more strictly are we warned against all chemical foods, and every man or woman who cares for bodily well-being should take the words of a celebrated American Chemist, who (with that instinct which so often in the case of true men of science seems to anticipate the deductions of strict science) has given this emphatic opinion :

> "I cannot but think, although it may be a prejudice, that chemicals had better be kept out of the kitchen." *

It is the Biologist and not the Chemist who gives the full reason. To the Chemist it is a "prejudice." He sees that Nature is very exact and exacting. That the smallest particle of matter out of place will spoil his experiment. Scrupulous cleanliness of flask and beaker, and tube and retort, and exactness of weighing, for a true chemical balance, will weigh the "small dust" of the best common balances, are his familiar requirements.

---

* Professor J. Cooke, junr. The full quotation is as follows :—
" When soda and cream of tartar are used in making bread, this salt (" Rochelle salt") remains in the loaf. The amount formed is too small to be injurious, but I cannot but think, although it may be a prejudice, that chemicals had better be kept out of the kitchen." *The New Chemistry*, p. 148.

Here we have a first disciplining of the mind for the still greater exactness which is needful in the sphere of vital phenomena.

For Chemistry is the preparatory school to Biology, and we must never forget these lessons in exactness, which even dead matter teaches. We have seen how tolerant the human body *seems* to be. How glibly we may talk of amounts " too small to be injurious." But has any one ever shown that Nature accepts the maxim, that *small quantities, just because they are small*, can be safely neglected ? " Too large to be injurious" would be, at least, a safer, if not a truer, maxim.

Alma has few restrictions, and none in the way of either the *cultured* palate or the *natural* taste, to place upon her guests. She is a liberal provider. Her " touch nots," " taste nots," are few, but they are extremely imperative and exact. When she says *taste* not, she means it, not a drop, not a grain, not a *soupçon*.

For certain acquired tastes, in every sense expensive, she has only scant tolerance. The " fragrant weed," as the writer can testify, loses its attractiveness under her regime. She plants a dislike for smoking, not in the *stomach*, but in the general fastidiousness of the smoker. He is insensibly weaned by finding that he cannot get tobacco or cigar good enough, at any price. Even a cigarette taken " in memoriam" of past pleasure, is somehow not finished. He drops it. It is not the same to him. He knows not why.

The smallest pinch of mineral salt in soup or vegetable, becomes suggestive of a handful thrown in by a jealous kitchen-maid to spite the cook. " *Tantaene animis,*" &c. ? As for soda, and all its tribe of baking-powders and self-raisers (which raise themselves and their inventors or vendors, at the expense of a foolish and suffering public), the nauseous alkali betrays itself at once, in a flavour suggestive of Dead-sea water, or the more familiar savour of a badly washed pocket-handkerchief.

Again the taste for flavours of disease and corruption, such as for the long-killed pheasant, the high venison, and even the idyllized *paté de foie gras*, simply vanishes of its own accord.

It is curious how Nature takes up this matter quite independently of reason. Of course no man of sense would care to pay *too dear* for the enjoyments of his palate; yet the writer knows of one poor man who would, and he fears does, pay any price for the truffle, and yet utterly despises the "goose's enlarged liver," and, unlike Virro in Juvenal's Satire,* would not care to have one set before him were it as large as the goose, and that as big a goose as himself.

Nay, he is a very *Alledius* in this matter, and I

---

* The lines referred to are as follows, *Juv. Sat.* v. 113.:

"Anseris ante ipsum magni jecur, anseribus par
Altilis, et flavi dignus ferro Meleagri
Fumat aper: post hunc raduntur tubera si ver
Tunc erit, et facient optata tonitrua coenas
Majores: tibi habe frumentum, Alledius inquit
O Libya, disjunge boves, dum tubera mittas.

Madan translates literally if not very intelligibly; "Before himself [Virro] is placed the liver of a great goose, equal to geese, a crammed fowl, and worthy the spear of a yellow (haired) Meleager, smokes a boar; after him truffles are scraped if then it be spring, and wished-for thunders make suppers greater: 'Have thy corn to thyself,' says Alledius, 'O Libya, unyoke your oxen, while you will send truffles.'

have heard him make a sort of stuttering pun (as he thinks in the style of Charles Lamb; he *is* something like Lamb in the stutter), and declare with solemnity,

"May I live to see the day when truffles shall be true tr-triffles in price."

If one remembers that all the most esteemed flavours, even now, come from the vegetable kingdom—that a clever French cook can make any one meat represent any other, and all this by the deft use of savours and flavours drawn chiefly, if not altogether, *from the vegetable kingdom*, we see there need not be such a Revolution in a Biological Reformation, as one would suppose. And assuming that this is a true interpretation of Nature and her laws, we may confidently affirm, supported by the analogy of every science and every discovery, that true Biological cookery, when once it is understood and practised, will give a *luxury* as far exceeding the present reach of cookery as the Pullman Railway car excels the pillion.

The writer may claim a fair acquaintance with what is considered *good living*. For many years before he became either a teetotaller or a vegetarian, he was at least an occasional guest at some of the most luxurious tables. Even an "*alderman's feast*," in a luxurious London company, at which he was a guest, had only the interest of curious custom, and nothing of *gastronomic novelty* to present to him.

He believes that twelve years' total abstinence from alcoholic drinks, and about five years' total abstinence from all flesh and non-organized mineral, have increased rather than diminished the sensitiveness of the gustatory and olfactory nerves, and enable him to have a keener and more discriminating enjoyment of flavours and odours than ever.

But whether my views are, or will be, borne out by the experience of others in this matter of luxury and gastronomic epicureanism or not, I confess is to me a matter of little moment. It is the bearing of these truths upon the food-supply of the millions, who want to get life and health, and not luxury, out of their sustenance, that seems to me all-important.

Now while fully granting the marvellous benefit which the glorious science of chemistry has been to the masses of mankind, it has, in one important point, been bitterly disappointing. I believe that a true Science of Biology is destined to fully perform all that Chemistry seemed to promise, or rather it will teach us that Nature is ever far kinder to us than in our ignorance we can believe.

To show what these hopes were, I refer the reader to Herschel's classical *Discourse on the Study of Natural Philosophy*. I shall give a sufficient quotation to make the point quite clear.

Herschel's foot-notes are inserted in brackets.

"The transformations of chemistry, by which we are enabled to convert the most apparently useless materials into important objects in the arts, are opening up to us every day sources of wealth and convenience of which former ages had no idea, and which have been pure gifts of science to man. Every department of art has felt their influence, and new instances are continually starting forth of the unlimited resources which this wonderful science develops in the most sterile parts of Nature. Not to mention the impulse which its progress has given to a host of other sciences, which will come more particularly under consideration in another part of this discourse, what strange and unexpected results has it not brought to light in its application to some of the most common objects? Who, for instance, would have conceived that linen rags were capable of producing *more than their own weight* of sugar by the simple agency of one of the cheapest and most abundant acids? [* The sulphuric. Bracconot, Annales de Chimie, vol. xii., p. 184]—that dry bones could be a magazine of nutriment, capable of preservation for years, and ready to yield up their sustenance in the form best adapted to the support of life on the application of that powerful agent, steam, which enters so largely into all our processes, or of an acid at once cheap and durable? [† D'Arcet, *Annales de l'Industrie*, Février, 1829]—that sawdust itself is susceptible of conversion into a substance bearing no remote analogy to bread; and though certainly less palatable than that of flour, yet no way disagreeable, and both wholesome and digestible, as well as highly nutritive? [‡ See Dr. Prout's account of the experiments of Professor Autenrieth of Tubingen, *Phil. Trans.*, 1827, p. 381. This discovery, which renders famine next to *impossible*, deserves a higher degree of celebrity than it has obtained.] What economy, in all processes where chemical agents are employed, is introduced by the exact knowledge of the proportions in which natural elements unite, and their mutual powers of displacing each other! What perfection in all the arts where fire is employed, either in its more violent applications (as, for instance, in the smelting of metals by the introduction of well-adapted fluxes, whereby we obtain the whole produce of the ore in its purest state), or in its milder forms, as in sugar-refining (the whole modern practice of which depends on a curious and delicate remark of a late eminent scientific chemist on the nice adjustment of temperature at which the crystallization of syrup takes place); and a thousand other arts which it would be tedious to enumerate!" (*Nat. Philosophy*, chap. iii., p. 64, Lardner's Edit. 1833.)

This was written more than sixty years ago. By simple lapse of time we are in a position to judge of those matters of which Herschel could then speak only as in prophecy. With consummate discernment did that great man "look into the seeds of time," and, with *almost* unerring correctness, "say which grain would grow and which would not." Indeed, chemistry has more than fulfilled all the promise of its youth, and the prophecies going before, with one, and one only, exception.

It has facilitated all arts, it has created new industries, it has brought a livelihood to thousands, and even renewed politically, and altered, the face of the globe. Its philosophy has affected all philosophy, and our deepest speculations about matter and spirit are chemical, at least in form. Chemical processes are now applied to star and nebula, on the very farthest confines of space, and Herschel's own magnificent and untiring labours, in cataloguing the constellations, has been handed on to chemistry to complete. For chemical photography is now being employed to map the whole heavens, and those wondrous eyes, supplied by chemical science, can behold things invisible to mortal sight. Even darkness is no darkness to them, and neither light nor heat is needed by the actinographic power, which faithfully records rays which have no correspondence with any of our senses.

What promise has not been kept, what prophecy has not been more than fulfilled? One promise, one prophecy, but that one the most practically important to mankind of all, for "all the labour of a man is for his mouth, and yet the appetite is not filled."*

Where is the food supply that was to make famine next to impossible? There is no lack of material—we have sawdust and vitriol in abundance, to fill all things living, at least all humanity. And starving men and women and children are not squeamish, and would eat a substance bearing no remote analogy to bread, and "both wholesome and digestible, as well as highly nutritive." And chemistry in all matters, *which require only the rearrangement of molecules,* has so far surpassed all expectation that not out of sawdust, but almost *any dust*, containing the chemical constituents of organic bodies, it would have found out many ways of manufacturing food.† Can we explain this? Most assuredly we can. The expectation was utterly un-

---

* The so-called "Chemical food" is simply a mixture (not in chemical combination) of the following drugs; Phosphate of lime, phosphate of iron, phosphates of soda and potassa, besides free phosphoric and hydrochloric acids. (See Macnamara's *Neligan's Medecines,* p. 777).

† The student will find the following quotation very interesting :—"We have also learnt, that owing to this identity of composition, many animals are saved the labour of forming these

reasonable; as absurd as to expect astronomy to put us in possession of property in the moon, say, the " three acres and the cow" that jumped over that luminary.

The very business of Chemistry is to *alter* and *re-arrange* the molecules of matter, and the proper condition of food for man is *that particular arrangement* in which sun and plant power have left them; nay, even too long delay will permit the subtle energy to escape, and none must rudely violate that shrine of the sun-god, which is concealed in the foliage of all vegetation, on pain of being misled by false oracles, from deceiving influences which then enter Apollo's deserted sanctuary.

For Chemistry is the King Midas of the sciences, and turns to gold whatever it touches; but in so doing it fatally unfits whatever it but touches for human food.

And just as in the old myth when Midas was appointed umpire in the contest between Pan and

proximate principles from their elements; and have only to rearrange them as their exigencies may require. The task of forming the proximate principles is thus left to the inferior animals, or to plants; which are endowed with the capacity of compounding these proximate principles from matters still lower in the scale of organisation than the animals and plants themselves. Hence there is a series, from the lowest being that derives its nourishment from carbon and carbonic acid, up to the most perfect animal existing: each individual of the series preferring to assimilate other individuals immediately below itself; but having on extraordinary occasions the power of assimilating all, not only below but above itself, in the system of organised creation." (Dr. Prout's *Stom. and Ren. Dis.*, Syd. Soc., p. 459.) Here we have in Prout's own words a splendid example of the two fatal defects of his learned and ingenious system:

   1. Classification of food, based on chemical composition.
   2. Almost total ignoring of the Law of Interchange.

Apollo, and rashly disregarding Apollo, was gifted with the ears of an ass, so do the presumptuous intrusion of chemical methods into cookery, and the chemical classification of food, when it leads to disregard of the Law of Interchange, turn a royal science into folly.

Not the recommendation of a Herschel, not the constructive genius of a Prout, or the splendid literary power of a Huxley, will avail to make man find that food wholesome, which is taken at second-hand, or those chemically correct, but biologically faulty, materials suffice for a healthy human body.

Chemistry has given us no medicine that can stop any disease, unless with the life of the patient, and that food is best which is farthest from the physician's prescription.

In the whole range of medical History it is utterly impossible to find more than one solitary remedy that has stood its ground for two centuries, and now retains universal acceptance.

Perhaps the longest-lived was the "Praised Liquid" which Sydenham valued,* the *Liquid Laudanum Sydenhami*: who will dare to praise *Laudanum* now? Or some may think of another drug which Sydenham valued highly, the Peruvian bark. Some value it still, but its course is nearly run, and I think I may fairly estimate that fully one-half of the medical men that issue from our schools would as soon drink laudanum, or any other poison, as take Peruvian bark or its derivative, quinine, *themselves;* and the whole body of Homœopathic Practitioners would to a man

* See Syd. *Op. Om.,* p. 174.

reject Sydenham's use of it. Calomel, I need not name, and I can think of no one remedy, that has universal suffrage to boast for two centuries, but lemon juice alone.*

We have already explained this biologically, that lemons are a veritable embodiment of sun and plant power. It does not come within the scope of this little work to go into any details about food or medicine. I content myself now with the assertion that the promise, made wrongly in the name of Chemistry, will be found to be fully kept in Biology.

And that, for a very small cost, far under the usual butcher's bill alone, can be purchased all that is needed to maintain a life of full mental and bodily vigour in the healthiest and most active condition. Even now Nature's abundant supply is brought to our very doors by commerce, and is ready (though in nothing like the quantity, and at far above the cost to which it must be reduced, when rational living becomes the rule and not the exception), for every one who knows *what* to seek for, and *where* to seek for it.

To show how much good food Nature may provide for us when we are starving ourselves *in*, or *by, imagination*, the following remark from Thackeray's *Irish Sketch Book* is instructive:

> "Here we saw the first public evidence of the distress of the country. There was no trade in the little place, and but few people to be

---

* In the year 1600, Commodore Lancaster sailed from England for the Cape of Good Hope. His men were kept quite free from scurvy by the administration of three table-spoonfuls of lemon juice every morning.

seen, except a crowd round a meal-shop where meal is distributed once a week by the neighbouring gentry. There must have been some hundreds of persons waiting about the doors; women for the most part: some of their children were to be found loitering about the bridge much further up the street; but it was curious to note, amongst these undeniable starving people, how healthy their looks were. Going a little further we saw women pulling weeds and nettles in the hedges, on which dismal sustenance the poor creatures live, having no bread, no potatoes, no work. Well! these women did not look thinner or more unhealthy than many a well-fed person. A company of English lawyers, now, look more cadaverous than these starving creatures." (*Sketch Book*, chap. II, p. 27.)

You see the nettle-protoplasm, being also vegetable, is excellent food. To quote Huxley, with a correction—"It appears to be a matter of no great moment *what plant* I lay under contribution for protoplasm."

The conditions under which plants yield us aliment are, in fact, simply these—(1.) There must be a *sufficiency* of materials, and, of course, of the right sort, and in *as far as possible* the right proportions. Nature can select out what she wants to some extent, but it saves much trouble and vital expense to give the right proportions (compare the saving in manufactures effected by a knowledge of chemical proportions).

(2.) The vegetable nutriment must be *get-at-able*,

without chemical process, which would destroy the plant-form.

Hence most poisonous plants are excluded from being sources of food, but not all. For example, tapioca comes from a highly poisonous plant, but then the poison is very volatile, (chiefly prussic acid), and only needs heat (and no chemicals) to drive it away.

All chemical processes, including even boiling, or maceration, in *hard water*, tends to injure food.

No one who has ever tried fairly the use of pure soft water would go back to hard. Vegetables keep their beautiful colours when boiled properly in soft water, and no cook would wish to colour up with salt or soda, who has fairly tried the pure water. By "water" I always, like all chemists, mean H_2O, not a solution of lime-salts, and other abominations.

Already a grand step in the right direction has been made by the "Salutaris" Water Company, which supplies distilled water, without aëration, at a cheap rate. But every house in our towns ought to have pipes supplying really pure water. And if we remember that by taking off the pressure of the air, water can be made to *boil at the freezing-point*, there can be no insuperable obstacle in the way of a cheap and abundant supply of evaporated and re-condensed water; once its dietic value is recognized. In the country, when sufficiently far from towns and factories to have pure air, the sky is, in our climate, an all-sufficient source of pure water, if a proper provision were made for its collection and storage, only needing to be boiled.

But, for fear of seeming to be an advocate of nettles and soft water only, as the Diet of Health, let me again remind the reader of the Bounty of Nature. I have spoken of nettles as staving off starvation, but Health is equally far from starvation as from gluttonous repletion in her Diet. In the beautiful words of the poet Spencer:

> "There Alma, like a virgin queene most bright,
> Doth florish in all beautie excellent;
> And to her guestes doth bounteous banket dight,
> Attempred goodly well for health and for delight."*

* *The Faery Queene,* Canto xi., 2.

CHAPTER IX.

THE IMPORTANCE OF USING NO INTRACTABLE MATERIALS FOR THE CONSTRUCTION OF THE HUMAN BODY.

WE all acknowledge the horrible evils which follow the use of fusel oil, or the maddening ingredient of new ardent spirits. Can we find a tincture that is not itself tinctured with fusel oil? Not one. As prepared according to the pharmacopoeia with rectified spirits diluted down to proof.

But our observation is not limited to the case of tinctures, nor is fusel oil our only enemy. Nothing of the drug-nature, nothing that stays as a secret enemy, or even lingers as a tiresome guest, has any business ever to enter our bodies.

Ancient Philosophy and modern science agree exactly on this one point—viz., that our bodies are, by natural law, in a state of *flux*. Like that of a mountain-torrent, our very existence, so far as regards our bodily life, depends upon the ceaseless continuance of the flow; when the stream fails our places know us no more.

Mother Nature is always an Alma Mater to those who keep her statutes and observe her laws, and even to the transgressors, though *their* way is indeed hard, *her* way is always to be doing the very best, most kind, wise, and healing thing she can, from moment to moment.

Do not force her to build your House of Life with

wrong, or defective, materials. Help Alma to repair her castle-walls, and above all aid her in the grand endeavour of her very being, to drive back and keep out all intruders. Then the Mystery of Pain will find in you that solution which, after all, is the *only satisfactory solution,* viz., that there should be " NO MORE PAIN."

Bodily pain is simply Alma's voice, rallying her forces, trooping her guard, or calling out for assistance. She keeps up her cry, often full of unspeakable agony when it is disregarded, or, more agonizing still, is misunderstood, until it is silenced in despair, or changed into the strain of victory—the voice of joy and health.

"What is your life? Ye are a vapour that appeareth for a little time, and then vanisheth away,"* is the demand and answer of a sacred writer, and modern chemistry vouches for the literal truth of the description as regards our bodies, for we are indeed principally water-dust. This is all as it should be, the watery part is, if anything, the most precious part of our atmosphere, and in its myriad manifestation is a power for good, which makes habitable the earth and glorifies the heavens. But a drugged body is like a smoke and poison-laden fog, and where there is an assemblage of them, it makes "a pestilent congregation of vapours."

How much of that strange restlessness, combined with a still more strange narrowness, and even crampedness of mind—like the fixed unfixedness of a monomaniac's ideas, which characterizes so much of what passes for scientific speculation—may be very easily explained as due to drugged brains.

"I cannot give you a wholesome answer," says

* St. James' Ep., chap. IV., v. 14, Rev. Vers.

Hamlet, "my wit's diseased." Many great actors on the stage of science now assume the part of Hamlet, and fearfully unwholesome are the answers we get.

Look at the death-bed of the late Emperor of Germany. Look at it, as all the world was lately *forced* to do, (however unwilling to enter into so sacred a place except with hushed and reverent sympathy) in its medical aspect.

See the representatives of the healing-science of this age, the best presumably that the world could offer, crowding round that Sufferer of needless agony, and thronging the torture-chamber made such by their own acts.

Frederick the Noble has died before his time, but he has neither lived nor died in vain. He lived to make, or help in making, his loved Fatherland great, united, strong, and free. He died, and not Germany only, but the world will be helped towards its deliverance from a bondage which shortens and embitters life, and makes death horrible.

The general truth, that a drugged organism cannot heal itself, has been taught us, in the fact, acknowledged by every honest medical man, that *no truly organic disease is curable.*

But have we not had more than enough of that wretched lesson? Cannot we now turn to the converse truth, that NO LIMIT CAN BE SET TO THE CURABILITY OF DISEASE—ORGANIC AS WELL AS FUNCTIONAL—IN AN UNDRUGGED AND RIGHTLY-FED BODY?

To this fact I can testify, from my own observation and practice. But once the fact is fully grasped, and when the study of these general truths, as illustrated throughout the living world, has woven them into the conceptions of things, then we shall realize something of the enormous mischiefs which the disregard of Nature's clear teaching has caused, and (happier and more useful result) the limitless advantages of attending to this teaching.

THE END.

ERRATUM.—Page 60, line 4 from top, for Béchat read Bichat.

15 York Street, Covent Garden,
London, *October* 1889.

Mr Redway's Publications

New and Forthcoming Works

Demy 8vo, white cloth, gilt, 5s.

In Tennyson Land:

BEING A BRIEF ACCOUNT OF THE HOME AND EARLY SURROUNDINGS OF THE POET LAUREATE, AND AN ATTEMPT TO IDENTIFY THE SCENES AND TRACE THE INFLUENCES OF LINCOLNSHIRE IN HIS WORKS.

BY J. CUMING WALTERS.

ILLUSTRATIONS BY F. G. KITTON.

CONTENTS:—Tennyson as an Artist—Specimen Pictures—His Range of Style—Love of England—An Early Effort—The Pleasures of Memory—Tennyson's Allusions to Cathedrals—General Aspect of Lincolnshire—A Night View—"The Dying Swan"—"The Gardener's Daughter"—"Locksley Hall: where is it?"—"Sixty Years After"—"The May Queen"—"The Lord of Burleigh"—"The Northern Farmer" and other Dialect Poems—Lincolnshire Types of Character—Country Sounds and Sights—Tennyson's Grand-parents—Louth Vicarage—The Poet's Boyhood—School-Life—His Brothers—Publication of "Poems by Two Brothers"—A Peep at the Original Manuscript—Tennyson's Remuneration—Lincolnshire Lanes—How Poems are Suggested—Familiar Sights—In the Poet's Land—Lincolnshire and the Seasons—Situation and Character of the hamlet—Arthur Hallam's Visits—The Rectory and the Lawn—Date of the Poet's Birth and Baptism—Mrs Tennyson—"The Owd Doctor"—Mournful Reminiscences—The Moated Grange—St Margaret's Church—"The Quiet Sense of Something Lost"—The Voice of the Brook—Its Course Traced—Katie Willows—"The Miller's Daughter"—"Maud"—The Poet's Affection for the Brook—The Nature of the Glen—"The Lover's Tale"—Scene of "Maud"—Influence of the Woods upon the Poet's Mind—Tennyson's Holiday Haunt—"The Lover's Bay"—Descriptions of the Sea—A Disillusion—Tennyson's Sympathetic Touch with Nature—Miss Jean Ingelow's Poems—Charles Dickens and Lincolnshire—Conclusion. Appendix:—Poems relating to Lincolnshire and Lincolnshire Character.

This work is choicely illustrated by drawings from photographs and sketches taken specially for the Author. The pictures include representations of—

SOMERSBY RECTORY.	THE BROOK.
THE MOATED GRANGE.	THE MILL.
LOUTH GRAMMAR SCHOOL.	
TENNYSON'S BIRTHPLACE.	

A few Large Paper copies, with the Illustrations printed on Japanese paper, may be had at special prices from the principal booksellers.

In Crown 8vo, Cloth, 7s. 6d.

Practical Heraldry;

Or, an Epitome of English Armory.

SHOWING

How, and by Whom Arms may be Borne or Acquired, How Pedigrees may be Traced, or Family Histories Ascertained.

By CHARLES WORTHY, Esq.,

Formerly of H.M. 82nd Regiment, and sometime Principal Assistant to the late Somerset Herald; Author of "Devonshire Parishes," &c., &c.

With 124 Illustrations from Designs by the Author.

Prospectus giving full contents may be had on application.

"A useful and compendious guide to the fascinating study of Heraldry. Orderly, lucid, and amply illustrated from designs by the Author. It justifies its claim to be a practical treatise."—*Notes and Queries.*

"Mr Worthy's Manual is addressed to the general reader, and explains the terms and rules of Heraldry in clear, non-technical language. It gives useful information about the sources of genealogies and the best methods of tracing them."—*Scotsman.*

"It was a happy thought of Mr Worthy to combine a treatise on Heraldry with an account of how to trace a Pedigree, and how to read an ancient record. Knowledge of the Science is to be obtained by the perusal, and such knowledge Mr Worthy is fully competent to give."—*Saturday Review.*

"Mr Worthy, known as the Author of Notes on 'Devonshire Parishes,' and who at one time assisted the late Somerset Herald, has issued a useful and practical work on a subject with which he is obviously well acquainted."—*Athenæum.*

"In addition to what is found in ordinary text books on the subject, Mr Worthy has some valuable notes on pedigrees and wills, with instructions as to how to trace a Pedigree."—*Court Circular.*

"Mr Worthy's exposition of the science of Heraldry is, on the whole, the best we know for clearness and compactness."—*The Beacon (Boston, Mass., U.S.A.)*

"Mr Worthy, in the volume just prepared, appears to have made a successful effort to compile a practical work containing information of interest to a large section of the community. The volume is well worthy of perusal; and his personal qualifications, he having been sometime principal Assistant to the late Somerset Herald, are such as to satisfy the reader of his general accuracy."—*Morning Post.*

"We have here a most useful book, and now that the study of Heraldry and the tracing of ancestry have become so general, a book which ought to be found in every gentleman's library. Mr Worthy is no mean authority on the subject, for in addition to a long and general practical experience, he held the position of principal Assistant to the late Somerset Herald."—*Western Antiquary.*

"The book takes the form of a handy volume of about 250 well printed pages, and is one that by arrangement and index is rendered easy for consultation."—*The Field.*

"Mr Worthy introduces a host of historical matter as to the origin of various coats of arms, seals, liveries, and the like, and by thus investing his subject with the elements of personal history, he has rendered his volume as interesting as it is useful."—*Court Journal.*

"Meets in a very efficient and satisfactory manner the long felt need of a simple, trustworthy, and readable treatise on the subject. Of Mr Worthy's qualifications for the task, nothing need be said : himself a member of a very ancient family, claiming descent from the Dukes of Normandy and from Charlemagne, he is a thorough master of his subject, and he may be accepted not only as a competent but a very agreeable Mentor."—*John Bull.*

"A lucid and very interesting introduction to one of the most fascinating of antiquarian sciences."—*The Sun* (New York).

---

2 *Vols. Demy 8vo, Cloth,* 25*s.*

The Philosophy of Mysticism

(*PHILOSOPHIE DER MYSTIK*).

By DR CARL DU PREL.

Translated from the German by C. C. Massey.

Contents :—Introduction—Science : Its Capability of Development—On the Scientific Importance of Dream—Dream a Dramatist—Somnambulism—Dream a Physician—The Faculty of Memory—The Monistic Doctrine of the Soul.

Extracts from a lengthy notice (over two columns) in the *Spectator*, Sept. 14 :—"The book, we may say at once, has been *thoroughly well translated* by Mr C. C. Massey—whose version of a good many passages we have compared pretty closely with the original—and the English reader will find in it *abundant subject for thought.* . . . Taken for what it is, and with the reserves already indicated, this TRULY ORIGINAL, TRULY ATTRACTIVE book may with a good conscience be recommended to an English public. Not the least of its merits is its fertility of suggestion of practical problems in introspective psychology—observations which the reader may make upon the one thing—namely, consciousness—which is always with him in waking hours, and of which he can infuse more than he perhaps supposes into the dim experiences of sleep."

"In the present work, which fills two sturdy volumes, he [Baron Du Prel] ventures into the region of dreams, and carefully and lucidly examines, as far as they can be examined by the light of science, the state of the dreamer, the somnambulist, and the clairvoyant."—*Pall Mall Gazette.*

"These volumes, admirably translated, are a most valuable addition to the bibliography of a subject which is now engaging the attention of both scientists and thinkers of all classes."—*American Bookseller.*

"We commend the book to all students of psychology. It should be added that the translation has been well done, and, unlike many from the German, is exceedingly readable, save in the more subtle philosophical portions, even to the ordinary reader."—*Nonconformist and Independent.*

"We could wish to follow up his fascinating speculations further in their bearings on the questions of immortality, ethics, and all the most important problems of the world. Du Prel's speculations differ from all others of the kind in this, that they purport to rest upon a basis of demonstrable facts. Whether these facts are genuine is a question worthy of more attention than it has hitherto received. He has the further advantage of being well equipped scientifically as well as metaphysically. Darwin and the scientists are as familiar to him as Kant, Schopenhauer, and Hartmann. Whether he has succeeded in pointing out a new path of psychological investigation it would be too soon to say definitely. But he has produced a very remarkable and striking book, and all who take an interest in these questions will do well to read it in Mr Massey's excellent translation."—*Literary World.*

"If the members of the Psychical Society have not already mastered the 'Philosophie der Mystik' of Baron Carl du Prel, doctor of philosophy, they have to hand a translation by C. C. Massey, in two volumes, which they can ill afford to neglect. This philosophic treatise on dream phenomena deals with Mysticism not as something unknowable, or isolated from experience, but as organically related with 'the totality of things.' The dream-life, our author undertakes to show, is as worthy of study as the waking life. . . . Dr du Prel's work teems with illustrations, derived from innumerable sources, of the wonders of somnambulism and clairvoyance, of the restorative virtue of the somnambulist's sleep, the health-prescriptions and cures of the clairvoyant, many of which must put the best physician to the blush. Of 'evidence,' in fact, there is more than sufficient to illustrate the dualism of consciousness, and to exercise alike the open-minded and the credulous."—*Saturday Review.*

"The exceedingly difficult and interesting questions relating to memory are discussed with great ability on the line of this double consciousness. The apparent permanence of all impressions—as shown by the reproduction of the most distant, complicated, and improbable in sleep or trance—the extraordinary gaps in the life of a *somnambule*, caused by the double state, are described most carefully."—*Scots Observer.*

"Mr C. C. Massey has sensibly enriched the student of transcendental philosophy by translating Du Prel's 'Philosophy of Mysticism.' . . . This translation, a piece of excellent work in a somewhat difficult field of labour, will be welcome to every one who is interested in the collateral development of German transcendental philosophy."—*Scotsman.*

"Speculations ingenious and far-reaching. . . . We thank Mr Massey for the general clearness of his rendering and for his lucid and persuasive introduction. . . . We cannot help feeling some interest and sympathy in the Baron's futile ingenuities and innocently boastful eclecticism: he is so bitter an opponent of narrow eighteenth century *Aufklärung*, and himself such a charming type of nineteenth century *Aufklärung*; he has read and misunderstood so much; he is so guilelessly persuaded he possesses the winnowing fan that can sift the true from the false in the beliefs of all ages; he is altogether so superior, so lucid and unbiassed an intellect, a pupil in all schools, and the judge of all."—*Athenæum.*

TENTH THOUSAND.   12mo, *Cloth*, 1s.

The Grammar of Palmistry.
By KATHARINE ST. HILL.
WITH EIGHTEEN ILLUSTRATIONS.

CONTENTS:—On the Outline and Mounts—On the Lines—On the Palm of the Hand and Lesser Lines—On Signs of Illness, Temper, and on Special Qualities—On Reading the Hands (*Examples*)—The Hands of Distinguished Persons—Glossary of Terms.

"The little manual is QUITE THE BEST THAT WE HAVE SEEN ON THE SUBJECT. . . . The expression of the soul through the body—and this is one method of it—is worth careful study."—*Light*.

"Those who provide themselves with the 'Grammar of Palmistry' will not require the services of a fortune-teller, but will be able—or may persuade themselves that they are able—by examining their own hands, after the manner prescribed in this little volume, to foretell their future fate. . . . The little book contains much interesting matter. The 'portraits' of the hands of several distinguished persons—the names of whom, however, are not given—are worthy of being studied. The text is illustrated by what may be termed descriptive drawings."—*Glasgow Herald*.

"The subject is one which is not without vogue in these days, and the little volume under notice enunciates clearly the principles of the science. The writer has endeavoured to disengage palmistry from the canons of necromancy and superstition with which old authors habitually mix it up. To enable the reader to grasp with greater facility the principles laid down, the book contains some twenty illustrations."—*Morning Post*.

In demy 8vo, Oriental Cloth, 10s. 6d.

The Indian Religions;
Or, Results of the Mysterious Buddhism.

CONCERNING THAT ALSO WHICH IS TO BE UNDERSTOOD IN THE DIVINITY OF FIRE.

By HARGRAVE JENNINGS,
AUTHOR OF THE "ROSICRUCIANS, THEIR RITES AND MYSTERIES," ETC.

This is probably the last work which will be published by that eminent Oriental scholar, Mr Hargrave Jennings, author of "The Rosicrucians." It is full of interest to those who study Buddhism, and also contains a store of curious learning on such matters as the following:—

HISTORY OF THE MAGI.
ASTRONOMY OF THE MIND.
SYMBOLISM AND THE SUPERNATURAL.
TEMPLARS AND THE FIRE PHILOSOPHY.
MAGNETIC SPECULATIONS.
SYMBOLISM OF COLOURS.
ROSICRUCIANS AND BUDDHISTS.
BRAHMINISM AND TRANSCENDENTALISM.
THEORY OF CASTE, &c., &c.

Analysis of Contents (pp. 8) may be had on application.

Demy 8vo, Cloth extra, 7s. 6d.

Bacon, Shakespeare, and the Rosicrucians.

By W. F. C. WIGSTON.

With Two Plates.

CONTENTS :—Chapter I.—John Heydon—The Rosicrucian Apologist—His Family—And Character—Identity of Bacon's "New Atlantis" with Heydon's "Land of the Rosicrucians"—Bacon's Hand to be traced in the famous Rosicrucian Manifestoes—Discovery of his Initials among the Members of the Fraternity—Proofs that the antedating of the Origins of the Rosicrucian Brotherhood was a Splendid Fraud. Chapter II.—The Prophecy of Paracelsus—A Stage Player one of the greatest impostors of his age, probably Shakespeare—Description of the Rosicrucian Manifestoes—Lord Bacon as Chancellor of Parnassus—Meeting of the Rosicrucians in 1646 at Warrington, at a Lodge, in order to carry out Lord Bacon's Ideas—Adoption of his Two Pillars, etc., etc.

"A most remarkable book. Like its predecessor, 'A New Study of Shakespeare,' one cannot open it without learning something. . . . But all the same the book is a curiosity, and NO SHAKESPEARE-BACON LIBRARY SHOULD BE WITHOUT IT."—*Shakspeariana (New York).*

"A noteworthy attempt has been made to fix the disputed authorship of the Shakespearian, and likewise of other writings, upon a set of literary eccentricities who existed in Shakespeare's time under the name of 'Rosicrucians,' after one Christian Rosenkreuz, a German noble of the fifteenth century. The fame of this curious literary 'sect' has just been revived by Mr W. F. C. Wigston. He endeavours to show that there existed in Shakespeare's day a learned college of men who wrote in secret, among whom were Lord Bacon, Sir Philip Sydney, Shakespeare, and Ben Jonson, and that these together concocted the plays."—*Westminster Review.*

"If Mr Donnelly's 'great cryptogram' should turn out to be a real discovery, we do not see why Mr Wigston's should not be so too. We fully believe that the two theories must stand or fall together."—*Notes and Queries.*

Opinion of James Hughan, author of many Masonic books, and reputed to be the highest Masonic authority in England:—"MY DEAR SIR,—I have carefully read your able article in the journal of the *Bacon Soc.* with great interest, and *much appreciation*. *Prima facie*, the case is made out, it appears to me, but beyond that I cannot go at present; but the evidence is so remarkable, as well as curious that no one of a thoughtful mind could possibly refuse your claim to consideration. The New Atlantis seems to be, and PROBABLY IS, THE KEY to the modern Rituals of Free-masonry. YOUR NOBLE VOLUME on Bacon, Shakespeare, and the Rosicrucians, does much to clear the way."

Crown 8vo, Cloth, 5s.

Problems of the Hidden Life.

Being Essays on the Ethics of Spiritual Evolution.

By PILGRIM.

Contents:—Dedication—An Aid to Right Thought—The Narrow Way—Orthodoxy and Occultism—The Goad of the Senses—Content and Satisfaction—Love's Aim and Object—The Two Pathways—Sir Philip Sidney—The Higher Carelessness—The Dark Night of the Soul—The Great Quest—Detachment—Meditation and Action—Death—Selflessness.

"We have no hesitation in saying these essays by an anonymous writer are thoughtfully written, and although, of course, we do not pretend to agree with the author's views, he states them with an earnestness and moderation which command our attention and respect."—*Literary World.*

"The book will be interesting to those who are acquainted with Indian philosophy. The student who cares only for the attainment of felicity among the Devas travels on the paths of 'Gnana,' 'Karma,' and 'Bhâhti.' Then follow rules for the 'Narrow Way.' The Christian Church is supposed to represent but 'one facet of the divine jewel of Truth,' compared with 'the all embracing Catholicity of the Occult Wisdom.'"—*Literary Churchman.*

12mo, Cloth, price 1s.

Handbook of Cartomancy,

Fortune-Telling, and Occult Divination.

Including Cagliostro's Mystic Alphabet of the Magi, the Golden Wheel of Fortune, and The Oracle of Human Destiny.

By GRAND ORIENT.

With 2 Plates.

The *St James' Gazette*, in an article entitled "Books on the Black Arts," says that "'Grand Orient,' in his 'Handbook of Cartomancy,' recommends a method of consulting the mystical wheel of Pythagoras which is apt to give very curious results."

"We have cheap science nowadays, cheap literature, cheap groceries, cheap everything, and why should not we have cheap magic as well? 'Grand Orient' at any rate thinks we should, and for the sum of one shilling has provided the public with a 'Handbook of Cartomancy, Fortune-telling, and Occult Divination,' which among other things lays bare the Oracle of Human Destiny, Cagliostro's Mystic Alphabet of the Magi, and the Golden Wheel of Fortune. By one or other of these methods the future may be made to yield up its secrets."—*Literary World.*

"A generous shilling's worth of amusement may easily be had out of the preternaturally solemn little volume."—*The Lantern.*

"The literature of Occultism, esoteric and practical, is now in the full flush of its renaissance. That literature has always been vast and widely distributed, although it has in the main been confined to the Latin tongues. English is the only non-Latin language which has any considerable body of books upon alchemy and astrology (to take the nobler arts of Occultism), and upon magic and divination among the lesser and baser of those arts. Mr Redway has for some years been the high priest who, as modern mystics would say, opens wide 'the door of the closed palace of the king.' He has just given us two more books in a department of Occultism which has of late become more vulgarised than any other. The secret arts of the diviner have been revealed to all the world. . . . Still there is a pleasing variety about 'Grand Orient's' little book which is very engaging. He will show you how to divine your future in dozens of different ways. But his chief reliance is in the Pythagorean Wheel, which is unluckily rather skittish."—*Scots Observer*.

12*mo, Cloth, price* 2*s*.

A Buddhist Catechism;

Or, Outline of the Doctrine of the Buddha Gotama, in the form of Question and Answer.

COMPILED FROM THE SACRED WRITINGS OF THE SOUTHERN BUDDHISTS FOR THE USE OF EUROPEANS, WITH EXPLANATORY NOTES.

By SUBHADRA BHIKSHU.

The Author thinks it has at length become incumbent on the Buddha's disciples to put forth a work suited to the intelligent appreciation of educated English readers, in which shall be set forth the sublime doctrine of the Buddha Gotama, not as a bygone system, but as a living source of pure truth accessible now to all who are athirst for spiritual knowledge.

Demy 8*vo, pp.* xi *and* 272, *Cloth,* 7*s.* 6*d.*

Gilds,

Their Origin, Constitution, Objects, and Later History.

By the Late CORNELIUS WALFORD, F.I.A., F.S.S., F.R.H.S., BARRISTER-AT-LAW.

Contains a Geographical Survey of the Gilds of Berks, Cambridge, Derby, Devon, Gloucester, Hants, Hereford, Kent, Lancashire, Lincoln, Middlesex, Norfolk, Northumberland, Oxford, Salop, Somerset, Warwick, Yorks.

About 500 pp., Demy 8vo, Cloth, price 18s.

The Development of Marriage and Kinship.

By C. STANILAND WAKE,
AUTHOR OF "SERPENT WORSHIP," ETC.

CONTENTS:—Preface. Introduction—Sexual Morality. Chapter I. Primeval Man. II. Supposed Promiscuity. III. Primitive Law of Marriage. IV. Group Marriage. V. Polyandry. VI. Polygyny. VII. Monandry. VIII. The Rule of Descent. IX. Kinship through Females. X. Kinship through Males. XI. Marriage by Capture. XII. Monogamy.

"The volume is a closely reasoned argument on a complicated and interesting subject, and will add to the reputation Mr Wake has already earned by his writings on anthropology. Portions of it have, we think, already appeared in English and foreign scientific journals and transactions, and this leads here and there to some repetition; but the work in its present form is consecutive and well arranged. It is easier reading than some earlier books on the same subject. . . . Mr Wake concludes his study of these difficult, but interesting questions by a chapter on modern civilized systems of monogamy; and on Christian ideas relating to marriage and celibacy."—*The Athenæum.*

"On the very complicated and unintelligible Australian marriage laws Mr Wake is well worth reading."—*Saturday Review.*

"A fund of valuable information in regard to savage usages all over the world. . . . Mr Wake gives a useful summary of the valuable investigation conducted by Mr Lorimer Fison and Mr Howitt into the Australian system of group-marriage."—*Literary World.*

"The supply of facts being so meagre, it is as a handsome contribution to those in regard to marriage and kinship that Mr Wake's present book is chiefly valuable. We say chiefly, because his deductions, to which the book naturally owes its interest, are given so guardedly and candidly, and with such full recognition of the necessity of further knowledge as to open the door to further inquiry rather than close it, as theories too often tend to do."—*Scots Observer.*

"Brimful of curious information; a work that all interested in genealogical questions will welcome, and which such as are not specialists will find much pleasure in studying."—REV. C. H. EVELYN WHITE in *The East Anglian or Notes and Queries, &c.*

"We shall not pretend to decide upon the correctness of any particular theory; but there need be no hesitation in saying that this work, in which sexual relations are considered in all their different forms of polyandry, polygyny, monandry, and monogamy, and the curious group marriages of the Australian aborigines and the Hawaiians, gives ample evidence that the author has made a thorough study of the subject in the light of the most recent researches, and has spared no pains in the collection of facts. His work is certainly a valuable contribution to the study of a very interesting and important subject."—*Scotsman.*

"Regarded as a mere storehouse of curious information as to the marriage customs which have at different times prevailed among different races, there is a great deal which is interesting in the volume before us."—*John Bull.*

Demy 8vo, pp. 315, *Cloth,* 10s. 6d.

Lives of Alchemystical Philosophers.

BASED ON MATERIALS COLLECTED IN 1815, AND SUPPLEMENTED BY RECENT RESEARCHES.

WITH A PHILOSOPHICAL DEMONSTRATION OF THE TRUE PRINCIPLES OF THE MAGNUM OPUS, OR GREAT WORK OF ALCHEMICAL RE-CONSTRUCTION, AND SOME ACCOUNT OF THE SPIRITUAL CHEMISTRY.

BY ARTHUR EDWARD WAITE.

TO WHICH IS ADDED A BIBLIOGRAPHY OF ALCHEMY AND HERMETIC PHILOSOPHY.

LIVES OF THE ALCHEMISTS :—Geber—Rhasis—Alfarabi—Avicenna—Morien—Albertus Magnus—Thomas Aquinas—Roger Bacon—Alain of Lisle—Raymond Lully—Arnold De Villanova—Jean De Meung—The Monk Ferarius—Pope John XXII.—Nicholas Flamel—Peter Bono—Johannes De Rupecissa—Basil Valentine—Isaac of Holland—Bernard Trévisan—John Fontaine—Thomas Norton—Thomas Dalton—Sir George Ripley—Picus De Mirandola—Paracelsus—Denis Zachaire—Berigard of Pisa—Thomas Charnock—Giovanni Braccesco—Leonardi Fioravanti—John Dee—Henry Khunrath—Michael Maier—Jacob Böhme—J. B. Van Helmont—Butler—Jean D'Espagnet—Alexander Sethon—Michael Sendivogius—Gustenhover—Busardier—Anonymous Adept—Albert Belin—Eirenæus Philalethes—Pierre Jean Fabre—John Frederick Helvetius—Guiseppe Francesco Borri—John Heydon—Lascaris—Delisle—John Hermann Obereit—Travels, Adventures, and Imprisonments of Joseph Balsamo.

" The chapter about Flamel is one of the most interesting in the book, but the longest and most enthralling is that containing a full account of the career of the infamous Cagliostro, whom Carlyle has immolated. This is really a romance of the highest interest. . . . There is abundance of interest in Mr Waite's pages for those who have any inclination for occult studies, and although he founds his work upon a book which was published in 1815 by an anonymous writer, yet he adds so much fresh matter that this is practically a new work. A valuable feature for students is the alphabetical catalogue which Mr Waite has prepared of all known works on hermetic philosophy and alchemy."—*Glasgow Herald.*

"Mr Waite has undoubtedly bestowed a vast amount of patient and laborious research upon the present work, inspired by the double conviction that the original alchemists had in fact anticipated and transcended the highest results of chemistry in the metallic kingdom, and had discovered in the twilight of the Middle Ages the future development of universal Evolution. The biographical sketches of the alchemists, both true and false, are curious reading, and the alphabetical catalogue of works on Hermetic Philosophy is surprisingly suggestive of ages when leisure was less scarce, and literature scarcer, than in modern days."—*Daily News.*

"The alchemists more generally known, such as Albertus Magnus, Roger Bacon, Raymond Lully, Flamel, Paracelsus, and Basil Valentine are dealt with fairly and fully, and the travels and adventures of Joseph Balsamo, alias Cagliostro, with his somewhat peculiar developments of Egyptian Freemasonry, are excellent and interesting reading. . . . Such an intelligent study of the subject must bring into relief the infinite possibilities which are contained in a combination of psychical insight with physical knowledge."—*Light*.

"The lives of the philosophers themselves are interesting and curious reading; the stories of Lully, Flamel, Valentine, Trevisan, and Zachaire are full of glimpses of mediæval times. To us, the most instructive and valuable of the lives is that of the prince of impostors, Joseph Balsamo, or Comte de Cagliostro, who died at the end of the last century."—*Spectator*.

"The old alchemists . . . may, however, with justice be regarded as the first experimentalists in analytical chemistry, and on this account are entitled to the gratitude of subsequent generations. The lives of the principal alchemists are briefly recorded, and their works mentioned. Amongst them are such familiar names as Thomas Aquinas, Roger Bacon, Paracelsus, Helvetius, and Delisle. The volume also contains an alphabetical catalogue of works on hermetic philosophy and alchemy."—*Morning Post*.

"A curiously interesting book which well deserves a place in the already extensive catalogue of remarkable books published by Mr Redway. Mr Waite has certainly not spared himself in the preparation and production of his work. . . . The narratives are in most cases romantic enough to interest the general reader, and will be more than acceptable to the mystic and occultist."—*Liverpool Daily Post*.

"A perfect storehouse of alchemystical lore. The lives of the principal alchemists are pleasantly and fluently told. . . . Then there is an essay on the true principles of the *magnum opus* of the alchemists, and an account of the so-called spiritual chemistry. Finally there is a bibliography of alchemy and hermetic philosophy. . . . There is doubtless something in Mr Waite's contention that modern psychical research tends to verify the alchemists' hypothesis of development in its extension to human intelligence. It is in accordance with the fitness of things that these ancient seekers after knowledge should have found in an age which is disposed to treat them with scant courtesy a conscientious, not to say enthusiastic, biographer, and apologist."—*Manchester Examiner*.

Price 6d.

Catalogue

Of a Portion of

The Valuable Library of the late Walter Moseley, Esq., of Buildwas Park, Shrewsbury, and other Important Books and Manuscripts

RELATING TO OCCULT PHILOSOPHY AND ARCHÆOLOGY; EMBRACING COLLECTIONS OF WORKS ON ASTROLOGY AND DIVINATION, SPIRITUALISM AND MESMERISM, ALCHEMY AND MAGIC, THEOSOPHY AND MYSTICISM, ANCIENT RELIGIONS AND MYTHOLOGY, FREEMASONRY AND THE ROSICRUCIAN MYSTERY, DEMONOLOGY AND WITCHCRAFT, GHOSTS AND VISIONS, IN THE ENGLISH, FRENCH, GERMAN, ITALIAN, AND LATIN TONGUES.

2 *Vols. Demy* 8*vo, pp.* 791, *Cloth, price* 21*s.*

The White King;

Or, Charles the First,

AND THE

MEN AND WOMEN, LIFE AND MANNERS, LITERATURE AND ART OF ENGLAND IN THE FIRST HALF OF THE 17TH CENTURY.

BY W. H. DAVENPORT ADAMS.

CONTENTS OF VOL. I.:—Personal History of Charles I.—Some of the Royal Children: Princess Elizabeth, Duke of Gloucester, Princess Mary, and Henrietta, Duchess of Orleans—The Court of Charles I.: Philip, Earl of Pembroke, The Countess of Carlisle, Sir Kenelm Digby—A King's Favourite: George Villiers, Duke of Buckingham—Notes—A Moderate Statesman: Lucius Cary, Lord Falkland—An Absolute Statesman: The Earl of Strafford—A Philosopher of the Reign of Charles I.: Edward, Lord Herbert of Cherbury—Glimpses of Life and Manners: The Strafford Letters—Appendix—Notes and Corrections—Index to Vol. I. CONTENTS OF VOL. II.:—Three Noble Ladies: Margaret, Duchess of Newcastle, Lady Anne Fanshawe, Mrs Hutchinson—The Arts in England during the Reign of Charles I.: 1. Music; 2, The Drama; 3. Painting and Architecture—Literature in the Reign of Charles I.: 1. The Courtly Poets; 2. The Serious Poets—Men of Letters in the Reign of Charles I.—Appendix—Notes and Corrections—Index to Vol. II.

"These two volumes belong to the gossip of history, they are essentially personal, and throw light upon much that is merely suggested in grave records. Mr Adams relates with vivacity, yet always with a careful regard of historical truth. . . . The scope of Mr Adams's work is comprehensive. He has carried it out with an intelligent thoroughness worthy of praise. Taken all in all, from the point of view of the general reader, his book is a satisfactory study in the intimate history of one of the gravest, yet also one of the most romantic cycles of our national life."—*Morning Post.*

"A peculiarly personal, and therefore interesting and readable book, while many of the pictures of social life and notable people are admirably vivid. Mr Adams has devoted special care to a narrative of the great Trial, and his chapters on the arts in England of that period are full of interest, those on the drama being quite worthy of preservation as a text-book for those seeking information of that particular kind. The author of 'The White King' has unquestionably done his work with a thoroughness which stamps it as a labour of love, and the two entertaining and instructive volumes are creditable alike to his industry and discrimination."—*Court Journal.*

"There is both judgment and eloquence in this story of the eventful life of the White King. . . . Like this popular writer's previous publications, this entertaining book is not meant to supersede history proper; it is rather an artistic clothing of the skeleton work of others, a graceful investing of dry details with circumstantiality, beauty, and realism. Nowhere is there to be found so ample and so faithful an account of the unfortunate Charles Stuart's doings and principles. . . . It needs a master hand, like Mr Davenport Adams, to evolve order from confusion, to define the one central figure, the King, around whom all, friends and foes, consciously or unconsciously revolved, and to demonstrate the influence and counter-influence of himself and his immediate surroundings on all England."—*Whitehall Review.*

SECOND EDITION.   *Crown 8vo, Cloth, price 6s.*

Dreams and Dream-Stories.

By ANNA BONUS KINGSFORD,

M.D. OF PARIS; PRESIDENT OF THE HERMETIC SOCIETY; PART AUTHOR OF "THE PERFECT WAY; OR, THE FINDING OF CHRIST.'

EDITED BY EDWARD MAITLAND.

"Charming stories, full of delicate pathos. . . . We put it down in wonderment at how much it outstrips our great, yet reasonable expectation, so excellent and noteworthy is it; a book to read and to think over."—*Vanity Fair.*

"Curious and fascinating to a degree . . . by certainly one of the most vivid dreamers, as she was one of the brightest minds, of her generation. . . . A curiously interesting volume."—*Court Journal.*

"Wonderfully fascinating . . . with invention enough for a dozen romances and subjects for any number of sermons."—*Inquirer.*

"More strange, weird, and striking than any imagined by novelist, playwright, or sensational writer . . . for the marvellous, the beautiful, and the vraisemblable, having Hawthorne's marvellous insight into the soul of things."—*Lucifer.*

"The preface is as singular as the stories themselves."—*Literary World.*

"All who knew Mrs Kingsford will remember that she was not only an Idealist, but an exceptionally gifted woman. . . . It is given to very few writers, even when broad awake, to tell such weird and striking stories in such lucid and admirable style."—*Lady's Pictorial.*

Crown 8vo, Cloth, 6s.

THE NEW AMERICAN NOVEL.

The Stalwarts;

Or, Who were to Blame?

By FRANCES MARIE NORTON,

THE ONLY SISTER OF CHARLES J. GUITEAU.

"The English reader will appreciate the excellent sketches of a settler's life in the far West, which form an important part of the book, and throw curious side lights on some phases of existence on the other side of the Atlantic."—*Morning Post.*

"The murder of President Lincoln and the plots and counterplots of American politics are interwoven with many bright and evidently faithful descriptions of life in the Eastern villages, the Western prairies, and the great cities of America. We cannot help wishing that the author had spared us the political incidents and contented herself with the family histories she relates so well. . . . A high tone pervades the book, but while the women, with but one exception, are self-sacrificing, devoted, pure and pious, the men are very poor creatures and in every way unworthy of their feminine belongings."—*Literary Churchman.*

SECOND EDITION. *Demy 8vo, about 500 pp., 8s. 6d.*

Christian Science Healing,

Its Principles and Practice, with full Explanations for Home Students.

By FRANCES LORD,

CO-TRANSLATOR OF FRÖBEL'S "MOTHER'S SONGS, GAMES AND STORIES."

CONTENTS:—The Twelve Lectures which usually constitute "A Course of Instruction in Christian Science"—A Simple Plan for Treatment (also arranged for use during six days)—General Directions on Healing—The Healer's Self-Training—Teaching—Books—Ought Christian Science Work ever to be paid for?—Home-Healing (Character and Conduct)—Circumstances—Children and Education—A simple Account of the Doctrine of Karma or Re-incarnation—A short Abstract of the Bhagavad Gita.

"There can only be one opinion about the work before me. A high moral tone, a lofty spirituality, a devout enthusiasm and large-hearted benevolence, are the characteristic features of the volume. I confess that in this age of materialism, gross and refined, it is refreshing to read a book, the supreme purpose of which is the bold assertion of the supremacy of spirit. Without professing to agree with all or even any of its conclusions, I have read the work with growing interest.

"The vital part of Christian Science appears to be the denial of sin and disease, as real entities in the world. And here the gifted authoress, I think, is perfectly right. Hell, sin, and disease have no Divine authority for their existence. They are the creations of man's fallen nature. There is but one life in the universe—*God*. The Christian scientist not only, however, denies the reality of sin and disease—which in a *certain sense*, is true—but goes a step further, and avers that they can be denied away—denied out of existence altogether.

"While I believe there is a considerable amount of error mixed with truth in this volume, I still think it will do great good in directing attention to the source of all disease, and in its continual insistence on a life of truth and purity. It only remains for me to say that it is written in a very charming style."—Rev. P. RAMAGE in *The Dawn, a New Church Home Journal.*

Crown 8vo, Cloth, with Frontispiece, price 6s.

Lesbia Newman.

A Novel.

By HENRY RORERT S. DALTON.

"*There is so much life in the book, it is sometimes so really clever, and it has such a fascination of audacity about it,* that one gets along over the vicious chapters in hope that they will be redeemed."—*Inquirer.*

Demy 8vo, Cloth, 5s.

The Influence of the Stars.
A Treatise on Astrology, Chiromancy, and Physiognomy.

By ROSA BAUGHAN.

To which is added a Treatise on the Astrological Significance of Moles on the Human Body.

Illustrated with a Facsimile of the Mystical Wheel of Pythagoras, and other Plates.

"Difficult as it may seem in this age of realism to attach any importance to what are generally considered accidents of birth, the fact that for many centuries every peculiarity of form or character was ascribed to astral influence by the most learned men of the time, may entitle the believers in astrology to an impartial hearing. The author of 'The Influence of the Stars' is evidently a firm believer in this and other occult sciences, and should she fail to convert her readers to her way of thinking the fault wil not be hers."—*Morning Post.*

"Full of wonder, mystery, and suggestion... Miss Baughan's volume is decidedly entertaining and instructive. Her researches have been deep, and she brings a mass of almost unique information into her pages for the reader to digest.... The book is got up in Mr Redway's well-known style, and is quaintly illustrated."—*Birmingham Daily Gazette.*

"Miss Baughan's book is distinctly interesting."—*Graphic.*

"Miss Baughan's book (Mr Redway should be praised for its not inelegant saffron binding) is not confined to palmistry. She has something to say about astrology and a good deal about physiognomy.... Upon chiromancy Miss Baughan discourses with the depth and subtlety which one expects from ladies when they take up with mysticism. Great high priestess of the art though she be, she cannot tell much about chiromancy that is not known to every haunter of tea-tables. But at least she makes her meaning clear, which is more than can be said of most feminine mystics."—*Scots Observer.*

BY A NEW WRITER.

Reggie Abbot:
An Historical Romance.

By NELSON PROWER.

ESSAYS IN THE LITERATURE OF ALCHEMY.

Small 4to, White Cloth, 10*s.* 6*d.*

The Magical Writings of Thomas Vaughan.
(*EUGENIUS PHILALETHES.*)

A Verbatim Reprint of his First Four Treatises: Anthroposophia Theomagica, Anima Magica Abscondita, Magia Adamica, The True Cœlum Terræ.

With the Latin Passages Translated into English, and with a Biographical Preface and Essay on the Esoteric Literature of Western Christendom.

By ARTHUR EDWARD WAITE.

"Some of Vaughan's reflections remind us of Jacob Boehm, but the Welsh mystic is, as a rule, more easily followed than his German brother. Indeed, with a few exceptions, the sense is clear enough to make the volume agreeable reading even to the uninitiated. . . . The seventeenth century was an age of plain speaking, and Vaughan, when differing from anyone, sometimes uses terms more forcible than elegant. Mr Waite supplies some interesting information about the history of occultism in his 'Introductory Lecture on the Esoteric Literature of the Middle Ages, and on the Underlying Principles of Theurgic Art and Practice in Western Christendom.'"—*Glasgow Herald.*

---

1 *vol., about* 10*s.*

Life and Writings of Jacob Behmen.

By DR FRANZ HARTMANN.

AUTHOR OF "LIFE AND WRITINGS OF PARACELSUS," ETC.

Writing only a few months ago, Mr Arthur Lillie said: "It is a pity Bohme's works are so scarce, for his philosophy, though clothed in somewhat obscure language, is really fine. Seen from the standpoint of Bohme, all the mythologies of the past become part of a vast science."

The publisher long ago sought to meet this want of a popular summary of Bohme's philosophy, but it was difficult to find anyone competent to undertake such a task. Dr Franz Hartmann however at length was persuaded to furnish a work which should be a pendant to his admirably succinct account of the teachings of Paracelsus.

THE ORIGINAL WORK ON PRACTICAL MAGIC.

Crown 4to, Cloth, Leather Back, Gilt Top, 25s.

The Key of Solomon the King.

(*CLAVICULA SALOMONIS.*)

Now first Translated and Edited from Ancient MSS.
in the British Museum,

By S. LIDDELL MACGREGOR MATHERS.
AUTHOR OF "THE KABBALAH UNVEILED," "THE TAROT," ETC.

With Plates.

This celebrated Ancient Magical work, the foundation and fountain head of much of the Ceremonial Magic of the Mediæval Occultists, has *never before been printed in English, nor yet, in its present form, in any other language,* but has remained buried and inaccessible to the general public for centuries. It is true that in the seventeenth century, a very curtailed and incomplete copy was printed in France, but that was far from being a reliable reproduction, owing to the paucity of the matter therein contained, the erroneous drawing of the Pentacles and Talismans, and the difficulty experienced at that time in obtaining reliable MSS. wherewith to collate it. There is a small work published in Italy bearing the title of the "Clavicola di Salomone Ridotta," but it is a very different book to this, and is little better than a collection of superstitious charms and receipts of Black Magic, besides bearing a suspicious resemblance both to the "Grimorium Verum," and the "Grimoire of Honorius."

Among other authors both Éliphas Lévi and Christian mention the "Key of Solomon" as a work of high authority, and the former especially refers to it repeatedly.

The Key of **Solomon** *gives full, clear, and concise instructions for Talismanic and Ceremonial Magic, as well as for performing various Evocations; and it is therefore invaluable to any student who wishes to make himself acquainted with the practical part of Occultism.*

Besides Seals, Sigils, and Magical Diagrams, nearly 50 Pentacles or Talismans are given in the Plates.

Crown 8vo, cloth, price 4s. 6d.

Paul of Tarsus.

By the Author of "Rabbi Jeshua."

"'Paul of Tarsus,' by the author of 'Rabbi Jeshua,' is a work of very considerable ability. . . . Literary facility, brilliancy of word-painting, wealth of what it is the fashion to call 'local colour,' this book undoubtedly possesses."—*Literary World.*

"The writer has carefully studied the history and characteristics of the time, and in an artistic, although very compressed form, and with GREAT LITERARY BEAUTY, he creates the historic surroundings and the atmosphere of his hero."—*Nonconformist and Independent.*

"Whoever the author of this work may be, and speculation has been rife as to whether it should be assigned to a distinguished Eastern explorer or to the late head-master of the City High Schools, it is certain that he is thoroughly at home in the details of Oriental life, and capable of presenting a life-like picture of the beginnings of Christianity stripped entirely of supernaturalism. The book as we say has *vraisemblance.* The writer carries us through the scenes of Paul's life and journeys, and fills up the background with such local colouring and scholarship that the readers are apt to forget how much is purely conjectural. . . . We commend the work, not as a contribution to the history of Paul of Tarsus, but as a picture of the times in which Christianity emerged."—*Freethinker.*

"This is a beautiful book. . . . It is a book of fascinating freshness and vigour. . . . It is most eloquently written, with great charm of style, and one devours it with that eager zest with which he devours a great imaginative work."—*Birmingham Daily Post.*

"Those who have read 'Rabbi Jeshua' will know what to expect in 'Paul of Tarsus,' from the pen of the same anonymous author. The work is most readable, though it is not at all like the popular biographies of the Apostle which appear in so great numbers. The authors of these are generally careful to show their erudition. The author of this work seems to be careful to hide his great and evident though it be. The justice of its local colour throughout, and the vividness of the pictures of Jerusalem, Antioch, and Rome, bespeak a scholar; while the charming style of the work, its simplicity and directness, show a writer of no mean literary skill."—*Scotsman.*

"A remarkable book. . . . The author has realised in his own mind a picture of Paul which, whether true or false, is vivid, and this he has reproduced in a style of unusual brilliance and power."—*Manchester Guardian.*

"The author has knowledge, imagination, and marked literary facility, and the result of these combined gifts is found in sketches which are rich in light, colour, life, and picturesqueness."—*Manchester Examiner.*

"Among those strange people who regard 'Robert Elsmere' as embodying in an attractive form the main teachings of Christianity, 'Paul of Tarsus' may find favour for its merely literary excellence, which is undeniable."—*Morning Post.*

"A considerable sensation was created seven or eight years ago by the publication of 'Rabbi Jeshua,' a brilliant rhetorical study of the life of Jesus by one who regarded Him as no Messiah, but as a pure-minded and high-souled enthusiast. The anonymous author now comes forward with a similar study. He fairly warns those 'whose hearts are firmly fixed in the lessons of their childhood,' and 'pious souls' who do not want their faith disturbed, to stop at the preface. . . . The great value of the work lies in its wonderfully vivid pictures of the social, religious, and political life of the times—pictures composed of skilfully grouped hints derived from a wide reading of contemporary, classical, and Talmudic literature."—*Christian World*.

"Such books as these, being diplomatic intermediaries between extreme agnostics and extreme dogmatists, can do nothing but good."—*Truth*.

A Magnificent Folio Edition of

Goethe's Faust.

From the German by John Anster, LL.D., with an Introduction by Burdett Mason.

Illustrated by FRANK M. GREGORY.

Mr Redway has the honour to announce the publication by him of the most magnificent edition of this immortal work yet produced.

The size is grand folio, 20½ by 15½ inches.

The text is by Dr John Anster, who was the earliest translator of *Faust* into English. His version, first published in 1835, gave pleasure to Coleridge, and is liked in Germany.

The illustrations (eighteen in black and white, ten in colour) form the special attraction of the volume. The charm of these illustrations is due hardly less to the artists who reproduced them than to Mr Frank Gregory, from whose wonderful drawings they were made. The new photo-aquarelle process has enabled us to embellish each copy of the book with a set of illustrations in colour, which an expert alone could determine were not actual water-colour paintings.

Mr Gregory, an American artist of undoubted genius, and secretary of the famous Salmagundi Club of New York, took up his residence in Germany in order to accomplish satisfactorily the work he had undertaken of illustrating *Faust*. He visited the scenes of Goethe's life and the supposed scenes of the Faust legend. His models of Marguerite, Mephisto and Faust were of course found in Germany; the elaborate costumes were kindly lent by the management of the Munich theatre, and all the accessories carefully supervised by those to whom *Faust* has been a life-long study.

The result is a splendid example of the bookmaker's art which should be in the hands of every connoisseur.

The entire edition (265 copies) has been produced in Germany, with the exception of the binding, which is the work of Messrs Burn & Co., London, from a striking design supplied by the artist.

1 vol., about 7s. 6d.

The Occult Sciences.

A CYCLOPÆDIA OF TRANSCENDENTAL DOCTRINE AND EXPERIMENT, IN FOUR PARTS,

EMBRACING CHAPTERS ON THEOSOPHY, MESMERISM, SPIRITISM, FAITH HEALING, THE MYSTICS, THE ROSICRUCIANS, THE FREEMASONS, DIVINATION, ASTROLOGY, AND ALCHEMY.

Written by Mr Waite, assisted by writers eminent in their own departments of study, this volume presents to the general reader an outline of every branch of occult science now studied. Facts are given profusely, opinions but sparingly. Vexed questions have been treated with respect for the views of experts who, equally eminent, differ. This single volume will serve many as well as a Library; others who wish to probe deeper into the mysteries of occult lore, will find themselves directed to those large and expensive works which the editor has throughout consulted.

---

612 *Pages, Large 8vo, with Plates,* 15s.

The Hidden Way across the Threshold;

Or, The Mystery which hath been hidden for Ages and from Generations.

AN EXPLANATION OF THE CONCEALED FORCES IN EVERY MAN TO OPEN THE TEMPLE OF THE SOUL, AND TO LEARN THE GUIDANCE OF THE UNSEEN HAND.

ILLUSTRATED AND MADE PLAIN WITH AS FEW OCCULT PHRASES AS POSSIBLE,

BY J. C. STREET, A.B.N.

This is a very extraordinary book. Its reputed author is known to be incapable of producing such a work, and the explanation of its appearance has been thus given by a lady who is well acquainted with him:—

"His book is a mere compilation of noble sayings, scrawled down out of books and *from the lips of the adepts* with whom he certainly has associated, and who have taught him some secrets."

This lady, one of the most cultured women of our time, has been content to become the pupil of an illiterate man, who, a few years ago, occupied a humble position in a drapers' shop, because he is known to be the medium of higher teachers. The book which bears his name has thus a peculiar interest for students of the "occult."

1 vol., about 7s. 6d.

A Walk from London to Fulham.

By the Late THOMAS CROFTON CROKER, F.S.A.

Revised and Enlarged by
G. W. REDWAY, F.R.Hist.S.

With nearly 200 Illustrations.

The copyright of Mr Croker's charming work, published by his son in 1860, having fallen into my hands, I have determined to reissue the book with such alterations in the text as the lapse of thirty years has rendered necessary; and with such large additions as will be involved by the extension of the 'Walk' to Hyde Park Corner and Knightsbridge, Kensington Gore and Old Brompton,—places Mr Croker did not visit in 1860. It will be my aim to eschew the 'dryasdust' element that so often prevails in works of this kind, and to **give** information that shall be absolutely trustworthy about objects of interest **by the way.**

THE LATEST WORK ON CHIROMANCY.
Crown 8vo, with 12 *plates,* **Cloth,** 2s. 6d.

Palmistry

And Its Practical Uses.

With Chapters on Astral Influences, and the Use of the Divining Rod.

Illustrations, Bible References, &c.

By LOUISE COTTON.

---

1 vol., about 2s. 6d.

Vegetarianism.

By Rev. JOHN H. N. NEVILL.

An attempt to popularize some of the teachings of Biology, and to show that abstinence from flesh-eating is really enforced by the teachings of generally accepted science.

Just published, small 4to, 350 pp., price 7s. 6d.

A NEW POSTHUMOUS WORK OF DR ANNA KINGSFORD,
FORMING A COMPANION BOOK TO
"THE PERFECT WAY."

Clothed with the Sun,

BEING THE BOOK OF THE ILLUMINATIONS OF
ANNA (BONUS) KINGSFORD.

WITH PREFACE, NOTES, AND APPENDIX, EXEGETICAL AND BIOGRAPHICAL.

EDITED BY EDWARD MAITLAND.

"A worthy companion to 'The Perfect Way' as a lasting monument to Mrs Kingsford's wonderful genius, great spirituality, and marvellous lucidity of insight into the 'hidden things of Nature and Religion.'"—*Lucifer*.

"Of surpassing interest to the psychologist, who, it seems to us, will regret that Mr Maitland has not taken advantage of the opportunity of prefixing to the present volume a biographical sketch of a singularly gifted woman. That Mrs Kingsford was a seer of the rarest lucidity and inspiration it would be easy to demonstrate. Rarely has the faculty of mental vision been so marvellously developed. Her book of 'Dreams and Dream-Stories' contains overwhelming evidence of some of the highest qualities of the poet; nor did she lack the power of adequate and beautiful expression. Of her in her character of prophetess and 'foremost herald of the dawning better age'—an age when the falsifications and corruptions of Christianity will have been replaced by the restoration of the great original truths of the primitive gospel—we must confess ourselves by no means qualified to speak. . . .

"Apart from these highest purposes which the volume is meant to serve, there is much that is interesting to an ordinary mortal, though, as has been indicated, it is in the main of a personal description. One is *intrigué* by Mrs Kingsford herself, rather than concerned about her doctrines; and yet some of these—see, for example, the very first chapter 'concerning the three veils between man and God'—are presented in so poetic and luminous a manner that, allowing always for varieties of interpretation, one cannot but be struck by their truth. . . . The appendix contains an account, all too short for the attention the subject awakens, of the overtures made to Mrs Kingsford and the editor of the late Laurence Oliphant as the representative of that arch-mystic Thomas Lake Harris. We have made no attempt to give any indication of Mrs Kingsford's views on the more serious subjects of which she speaks; they can be properly learned from the volume alone."—*Glasgow Herald*.

"The pure, sublime, and raptly abstract Anna Kingsford, being dead, yet speaketh. The pale, thin lady, the recondite student, the illumined seer, who yet occasionally on public platforms grasped the problems of exoteric life by the horns, has been gathered to her rest, the excalibar blade of the spirit having worn out its somatic sheath; but the visions that came to her by day and the dreams that visited her by night are by a loving hand unfolded before us. Happy was this gentle seer of visions and dreamer of dreams that her kindred spirit, Edward Maitland, survived her to perform her literary obsequies with affection and fidelity. . . . Not merely is it claimed for the writer of this extraordinary volume that she beheld with supersensuous vision the arcana covered by the timal conception of the Now, but that the Past lay before her as an open book, and that on its pages she could trace clearly the evolutionary history of her own previous existence on all the plains of purgatorial Karma. Such a claim was, perhaps, never before so gravely made by any human being. A writer who claims to definitely trace back the egoism of her own ego through all the countless æons of cosmogenesis, at the bare contemplation of which the brain absolutely reels, is certainly endowed with faculties possibly denied to every other individual of the human race."—*Agnostic Journal.*

Vols I. and II., 4to, Cloth, 21s. each. Vol. III. in preparation.
Subscribers' names are now being received.

Devonshire Parishes.

By CHARLES WORTHY, Esq.

" A very painstaking and pleasant volume [Vol. I.] which will be read with great interest by the topographer and genealogist."—*Vanity Fair.*

" In this volume [Vol. II.] Mr Worthy has given us the rest of his account of certain parishes in the Archdeaconry of Totnes, and the work, as a whole, forms a respectable addition to the number of our local histories. Records of this kind are often the means of ensuring the preservation of valuable objects. Mr Worthy has devoted considerable space to tracing the descents of manors and to the genealogies of the families which held them."—*Saturday Review,* July 6th, 1889.

With 8 illustrations, Cloth, price 7s. 6d.

The Light of Egypt;

Or, the Science of the Soul and the Stars.

This anonymous work is of American origin. It has been the subject of some controversy owing to the fact that it contains all the teaching which its author formerly imparted to pupils for a fee of 100 dollars. The pupils now complain that it is placed before the public for a few shillings. The author alleges that he has felt bound to try and check, by the publication of this book, the spread of "the subtle, delusive dogmas of Karma and Re-incarnation as taught by the sacerdotalisms of the decaying Orient."

Crown 8vo, pp. viii. *and* 446, *Cloth extra,* 7s. 6d.

The Real History of the Rosicrucians.

Founded on their Own Manifestoes, and on Facts and Documents Collected from the Writings of Initiated Brethren.

By ARTHUR EDWARD WAITE.

With Illustrations.

Contents:—Mystical Philosophy in Germany—The Universal Reformation—Fama Fraternitatis—Confession of Rosicrucian Fraternity—Marriage of Christian Rosencreutz—Rosicrucianism, Alchemy, and Magic—The Case of Johann Valentin Andreas—Progress of Rosicrucianism in Germany—Rosicrucian Apologists: Michael Maier, Robert Fludd, Thomas Vaughan, John Heydon—Rosicrucianism in France—Rosicrucians and Freemasons—Modern Rosicrucian Societies, &c.

"We desire to speak of Mr Waite's work with the greatest respect on the points of honesty, impartiality, and sound scholarship. Mr Waite has given, for the first time, the documents with which Rosicrucianism has been connected *in extenso.*"—*Literary World.*

"There is something mysterious and fascinating about the history of the Virgin Fraternity of the Rose."—*Saturday Review.*

"A curious and interesting story of the doings of a mysterious association in times when people were more ready to believe in supernatural phenomena than the highly-educated, matter-of-fact people of to-day."—*Morning Post.*

". . . The work not only of a refined scholar, but of A MAN WHO KNOWS WHAT HE IS WRITING ABOUT, and that is a great deal more than can be said for other books on the same topic. . . . Much that he has to tell us has the double merit of being not only true, but new."—*John Bull.*

"Mr Waite's book on 'Rosicrucianism' is a perfect contrast to the one which we noticed a month or two back. The latter is a farrago of ill-digested learning and groundless fancies, while the former is, at all events, an honest attempt to discover the truth about the Society of the Rosy Cross. . . . The study of 'Occultism' is so popular just now that all books bearing on such topics are eagerly read; and it is a comfort to find one writer who is not ashamed to confess his ignorance after telling us all he can discover."—*Westminster Review.*

"Mr Waite is A GREAT AUTHORITY on esoteric science and its literature. Those who have read his extremely interesting work upon the writings of Eliphas Levi, the modern magician, will expect in his 'History of the Rosicrucians' a treatise of more than ephemeral importance, and they will not be disappointed. . . ."—*Morning Post.*

"Some of the most interesting chapters in the book are devoted to an account of the four great apologists for Rosicrucianism: Robert Fludd, Michael Maier, Thomas Vaughan, and John Heydon. Each of these chapters contains much curious matter, very metaphysical and very transcendental, but worth being studied by those who appreciate the influence which the many forms of occultism have exercised upon civilisation."—*St James's Gazette.*

"To many readers the most fascinating pages in 'The Real History of the Rosicrucians' will be those in which the author reprints Foxcroft's 1690 translation of *The Chymical Wedding of Christian Rosencreutz*, which had originally appeared in German in 1616. THIS STRANGE ROMANCE IS FULL OF WONDERFUL THINGS."—*Saturday Review.*

"We would recommend Mr Waite's very painstaking volume to all who may be desirous to get to the back of the Rosicrucian mystery. . . . So much nonsense has been talked and written about this imaginary order that it is quite refreshing to find a writer competent and willing to reduce the legend to its true proportions, and show how and when it had its origin."—*Knowledge.*

"We have rarely seen a work of this description that was so free from all attempts at the distortion of facts to dovetail with a preconceived . . . . His style is perspicuous. . . . The most interesting portions of the book are those where the author is willing to speak himself. . . . To those students of occultism, whose palates, undebauched by the intellectual *hashish* of the rhapsodies of mysticism and the jargon of the Kabala, can still appreciate a plain historical statement of facts, we gladly commend the book."—*Nature.*

"'The Real History of the Rosicrucians' is a very learned book that will be read with deep interest by every one who has the slightest knowledge of the subject."—*Court Journal.*

"MR WAITE'S PAINSTAKING AND WELL WRITTEN BOOK IS ONE TO BE THANKFUL FOR. . . . The subject has too long (and never more than at the present) been the property of pseudo-learned mystery-mongers. . . . But scant justice can be done to a book like Mr Waite's in a short notice such as this, and therefore all that remains possible is to draw the attention of all interested in such literature to the careful chapters on the English mystics —Fludd, Vaughan, and Heydon—and to emphasise the estimate with which we commenced."—*Manchester Examiner.*

"There was need of a clear and reliable book on the subject. This need Mr Waite has supplied. He is a cultured writer, and has mastered the entire literature of his subject, the most of which is in the German language. His 'Real History' cannot fail to interest any curious reader. . . . The author is not a Freemason, and speaks slightingly of our fraternity; but he has undoubtedly produced THE MOST RELIABLE BOOK which has yet appeared in the English language on Rosicrucianism, and it will deservedly attract the attention of all scholars and curious readers who are interested in the subject."—*Keystone* (New York).

"Mr Waite has done an excellent service in reprinting in this handsome volume translations of the chief documents bearing on the secrets of the Rosy Cross."—*Literary World* (Boston).

"Mr Waite is not a trader upon the ignorance and curiosity of readers. . . . His own book is simply the result of conscientious researches, whereby he succeeded in discovering several unknown tracts and manuscripts in the library of the British Museum; and these, with other important and available facts and documents, . . . he now publishes, summarised or *in extenso*, according to their value, and thus offers for the first time in the literature of the subject, THE ROSICRUCIANS REPRESENTED BY THEMSELVES."—*Philadelphia Press.*

3 *vols. Crown 8vo, Cloth*, 6*s. per vol.*, SOLD SEPARATELY.

Dreamland and Ghostland:

An Original Collection of Tales and Warnings from the Borderland of Substance and Shadow.

EMBRACING REMARKABLE DREAMS, PRESENTIMENTS, AND COINCIDENCES, RECORDS OF SINGULAR PERSONAL EXPERIENCE BY VARIOUS WRITERS, STARTLING STORIES FROM INDIVIDUAL AND FAMILY HISTORY, MYSTERIOUS HINTS FROM THE LIPS OF LIVING NARRATORS, AND PSYCHOLOGICAL STUDIES, GRAVE AND GAY.

"It is a remarkable fact that men and women do like ghost stories. They enjoy being thrilled, and many of them read with avidity tales which deal with things out of the ordinary physical ken. IN THESE THREE VOLUMES THEY MAY SUP FULL OF THESE DELIGHTS."—*Scotsman.*

"There is plenty of amusing reading of this sort to be found in these volumes, both for believers and disbelievers in the supernatural."—*Court Journal.*

"Volumes which will test the credulity of the reader to the utmost, and the commencement of one of the stories might very well have served for the motto for the whole collection: 'It is almost useless to tell you the story, because I know you will not believe it.' We do not say for a moment that we disbelieve all the stories told here."—*Court Circular.*

"The psychological student would be wise to exercise a certain amount of caution. The general reader who likes ghost-stories and dream-stories for their own sake, in the straightforward old fashion, will find plenty of entertainment in these three volumes, and, thanks to the variety of sources from which the contents are drawn, no sort of monotony."—*Graphic.*

"The great novelty of the work is that the author has so arranged and trimmed the chain of narratives as to make them read like a three volume novel. . . . In truth, it is a novel in which the characters tell their own stories in their own way, and in their own language."—*Christian Union.*

"SHOULD BE SPECIALLY RELISHED THESE WINTER NIGHTS."—*The World.*

"Stories of the weird and eerie complexion which so many like to cultivate of a winter's night."—*Globe.*

"There is nothing that is in any way unhealthy in character. Those, therefore, who have a taste for the mysterious and the curious will find in 'Dreamland and Ghostland' A REAL TREAT. The narratives are at once both grave and gay, with touches of strangeness as to miraculous incidents and supernatural occurrences. But from first to last there is a rationalism as well as a piquancy in the records that make them instructive reading. Indeed, we believe that THERE IS NOT A BETTER WORK OF ITS KIND, so varied, so enchanting, and so well edited; or one that may be read with such profit."—*Christian Union.*

Large Crown 8vo, the Cover emblazoned and floriated with Stars and Serpents and Sunflowers, and the Arms of France and of Navarre. Gilt top, 10s. 6d.

The Fortunate Lovers.

Twenty-seven Novels of the Queen of Navarre.

TRANSLATED FROM THE ORIGINAL FRENCH BY
ARTHUR MACHEN.

EDITED AND SELECTED FROM THE "HEPTAMERON," WITH NOTES, PEDIGREES, AND AN INTRODUCTION, BY
A. MARY F. ROBINSON.

WITH ORIGINAL ETCHING BY G. P. JACOMB HOOD.

"After Boccaccio's, these stories are perhaps the best of their kind."—*Scotsman.*

"Miss Robinson's notes, and more especially her ably written introduction, which is practically a biography of Margaret of Angoulême, will enable readers to appreciate the 'personalities' in the stories more keenly than would otherwise be possible."—*Scotsman.*

"These tales of old-world gallantry cruelly depict certain phases of the life of an age as brilliant as it was corrupt, and must ever prove attractive to the antiquarian and the scholar. Mr Machen well preserves the incisive and quaint tone of the original text."—*Morning Post.*

"A REALLY CHARMING WORK OF ART AND OF LITERATURE."—*Athenæum.*

"Super-realistic as the love-stories now and then are, according to our notions of modesty, they have, one and all, a wholesome moral, and go far to throw light on an interesting period in the history of France. Handsomely bound and 'got up,' and furnished with a charming etching by Mr Jacomb Hood as frontispiece, the volume may well be recommended to all readers, and particularly to all students of history."—*Pall Mall Gazette.*

"The 'Heptameron' is itself, and independent of externals, an exceedingly pretty book, . . . a book of interesting and rather puzzling authorship, and lastly, one which strikes the key-note of a certain time better almost than any other single work."—*Athenæum.*

"No reader can resist the charm of these old-world stories. . . . Miss Robinson has exercised a sound and judicious discretion . . . without sacrificing too much of the large utterance and the rich aroma of the originals."—*Daily News.*

"The book may be recommended to all who wish to understand that singular mixture of piety and voluptuousness which distinguishes the French Renaissance."—*Athenæum.*

"The book is not quite one for indiscriminate presentation, but it is exceedingly well done, and is beautifully printed and bound."—*Glasgow Herald.*

"We owe her [Miss Robinson] thanks for having put in a worthy form before a new public a work to a great extent forgotten, and most assuredly not deserving forgetfulness."—*Athenæum*.

"Nothing can be better than the introductory chapter, and the notes and genealogical tables show that care for minute accuracy which is the fashion of the present day, and a very good fashion too."—*Westminster Review*.

"A book that people who like to saunter along the by-paths of history will revel in. As, at the present time, there are thousands of people who only care to read the gossip and scandal in 'society journals,' so there are readers of history who chiefly delight in the gossip and scandal of bygone days. From such people 'The Fortunate Lovers' is certain to meet with a hearty welcome, while even the more serious students of history will rise from its perusal with a fuller and better knowledge of the times it deals with."—*Literary World*.

"Many of the stories are not particularly edifying. . . . Has a distinct value as a contribution to historical literature."—*Court Circular*.

Crown 8vo, pp. viii. and 260, Cloth gilt, 6s.

Charles Dickens and the Stage.

A Record of his Connection with the Drama as Playwright and Critic.

By T. Edgar Pemberton.

With New Portraits, in Character, of Miss Jennie Lee, Mr Irving, and Mr Toole.

Contents:—The Stage in his Novels—Dickens as a Dramatist—Dickens as an Actor—Adaptations and Impersonations—The Stage in his Speeches—The Stage in his Letters—Dickens as a Dramatic Critic.

"The book is readable, as anything about Dickens is sure to be."—*Scotsman*.

"A charming work. Mr Pemberton has spared no pains to look up all sorts of details, and has added a full and excellent index."—*Birmingham Post*.

"He has done his work so completely that he has left little or nothing for anyone who should desire to follow in his steps."—*Literary World*.

"Brimful of anecdote and reminiscences of a generation now passing away, the book is stimulating as well as useful."—*Publisher's Circular*.

"An example of book-making that will not be viewed with disfavour by lovers of Dickens. . . . The book shows diligent research in many directions."—*Saturday Review*.

Crown 8vo, pp. xiv. *and* 360, *Cloth,* 7s. 6d.

Posthumous Humanity;

A Study of Phantoms.

By ADOLPHE D'ASSIER,

MEMBER OF THE BORDEAUX ACADEMY OF SCIENCE.

Translated and Annotated by Henry S. Olcott, President
of the Theosophical Society.

Contents :—Facts Establishing the Existence of the Posthumous Personality in Man—Its Various Modes of Manifestation—Facts Establishing the Existence of a Second Personality in the Living Man—Its Various Modes of Manifestation—Facts Establishing the Existence of the Personality in Animals, and concerning a Posthumous Animality—Fluidic Form of Vegetables—Fluidic Form of Gross Bodies—Character of the Posthumous Being—Its Physical Constitution—Its Aversion to Light—Its Reservoir of Living Force—Its Ballistic—The Nervous Fluid—Electric Animals—Electric Persons—Electric Plants—The Mesmeric Ether and the Personality which it Engenders—The Somnambule—The Sleep-talker—The Seer—The Turning-table—The Talking-table—The Medium—Miracles of the Ecstatics—Prodigy of Magic—The Incubus—The Obsessing Spirit—Causes of the Rarity of the Living Phantom—Causes of the Rarity of the Trans-sepulchral Phantom—Resemblance of the Spiritistic Phenomena to the Phenomena of the Posthumous Order—Lycanthropy—Glance at the Fauna of the Shades—Their Pre-occupations—How they Prolong their Existence—The Posthumous Vampire.

Truth says :—"If you care for GHOST STORIES, DULY ACCREDITED, EXCELLENTLY TOLD, AND SCIENTIFICALLY EXPLAINED, you should read the translation by Colonel Olcott of M. Adolphe d'Assier's 'Posthumous Humanity,' a study of phantoms. There is no dogmatism so dogged and offensive as that of the professed sceptic—of the scientific sceptic especially—who *ex vi termini* ought to keep the doors of his mind hospitably open ; and it is refreshing, therefore, to find such scientists as Wallace, Crookes, and M. d'Assier, who is a Positivist, in the ranks of the Psychical Research host. For my own part, though I have attended the seance of a celebrated London medium, and there convinced myself beyond all doubt of his imposture, I no more think that the detection of a medium fraud disposes of the whole question of ghosts, &c., than that the detection of an atheist priest disposes of the whole question of Christianity. Whatever view you take of this controversy, however, I can promise you that you will find the book interesting at least if not convincing."

"This collection of hopeless trash . . . Col. Olcott's notes are beneath contempt . . . a more piteous literary exhibition than the entire volume has rarely come under our notice."—*Knowledge* [?].

" An interesting and suggestive volume."—*New York Tribune.*

" The book is written with evident sincerity."—*Literary World.*

" There is no end to the wonderful stories in this book."—*Court Circular.*

"The book may be recommended to the attention of the marines."—*Scotsman.*

"A book which will be found very fascinating by all except those persons who have neither interest nor belief for anything but what they can understand."—*Manchester Examiner.*

"The subject is treated BRILLIANTLY, ENTERTAININGLY, AND SCIENTIFICALLY."—*New York Com. Advertiser.*

"Though this is a good deal to say, Mr George Redway has hardly published a more curious book."—*Glasgow Herald.*

"The ghostly will find much comfort in the book."—*Saturday Review.*

"The book has an interest as evidence of that study of the occult which is again becoming in a certain degree fashionable."—*Manchester Guardian.*

Demy 8vo, pp. xiv. and 307, Cloth, 7s. 6d.

The Life, Times, and Writings of Thomas Cranmer, D.D.,

The First Reforming Archbishop of Canterbury.

By CHARLES HASTINGS COLLETTE.

DEDICATED TO EDWARD WHITE, 93RD ARCHBISHOP OF CANTERBURY.

CONTENTS:—Cranmer at the University of Cambridge—Cranmer's Participation in the Proceedings of the Divorce of Henry VIII. from Catherine—His Second Marriage as a Priest—His Oaths on Consecration as an Archbishop—The Fate of Anne Boleyn: Henry's Marriages with Jane Seymour, Anne of Cleves, Catherine Howard, and Catherine Parr, and Cranmer's alleged Participation in these Acts—Henry VIII.'s Political and Social Reforms under Cranmer's alleged Guidance—Persecutions, and Cranmer's alleged Participation in them—The Progress of the Reformation under Henry VIII. and Edward VI.—Cranmer's Fall and Martyrdom—His alleged Recantations—His Writings—John Fox, the Martyrologist—The Beatification of Bishop Fisher, the Chancellor More, and others, as Martyrs.

"Mr Collette brings to his task both breadth and depth of knowledge, and a desire to be scrupulously free from prejudice."—*Globe.*

"He is animated by an anti-Papal spirit. . . . nevertheless, his book is readable."—*Scotsman.*

"No future student can afford to neglect his work."—*British and Colonial Printer.*

"His book deserves to be read, and his pleadings should be well considered."—*Anglican Church Magazine.*

"HE HAS STATED HIS EVIDENCE WITH A FULNESS AND FAIRNESS BEYOND CAVIL."—*Daily News.*

"Mr Collette avoids bitterness in his defence, and does not scruple to blame Cranmer when he thinks blame is deserved."—*Glasgow Herald.*

"On the whole, we think that we have in this book a just and impartial character of Cranmer."—*Record.*

"This book is a valuable contribution to the literature concerning a period which to the lover of religious liberty is of the deepest interest. . . . it is a work of research of learning, of sound and generally of impartial judgment."—*Rock.*

Post 8vo, with Plates, pp. viii. and 359, Cloth gilt, 10s. 6d.

KABBALA DENUDATA,

The Kabbalah Unveiled.

CONTAINING THE FOLLOWING BOOKS OF THE ZOHAR:—

1. THE BOOK OF CONCEALED MYSTERY.
2. THE GREATER HOLY ASSEMBLY.
3. THE LESSER HOLY ASSEMBLY.

TRANSLATED INTO ENGLISH FROM THE LATIN VERSION OF KNORR VON ROSENROTH, AND COLLATED WITH THE ORIGINAL CHALDEE AND HEBREW TEXT,

BY S. L. MACGREGOR MATHERS.

The Bible, which has been probably more misconstrued than any other book ever written, contains numberless obscure and mysterious passages which are utterly unintelligible without some key wherewith to unlock their meaning. *That key is given in the Kabbala.*

"A TRANSLATION WHICH LEAVES NOTHING TO BE DESIRED."—*Saturday Review.*

"Mr Mathers has done his work with critical closeness and care, and has presented us with a book which will probably be welcomed by many students. In printing and binding the volume is all that could be desired, and the diagrams are very carefully drawn, and are calculated to be very useful to all who are interested in the subject."—*Nonconformist.*

"We may add that it is worthy of perusal by all who, as students of psychology, care to trace the struggles of the human mind, and to note its passage from animalism through mysticism to the clearness of logical light."—*Knowledge.*

"Mr Mathers is certainly a great Kabbalist, if not the greatest of our time."—*Athenæum.*

The Kabbalah is described by Dr GINSBURG as "a system of religious philosophy, or more properly of theosophy, which has not only exercised for hundreds of years an extraordinary influence on the mental development of so shrewd a people as the Jews, but has captivated the minds of some of the greatest thinkers in Christendom in the 16th and 17th centuries." He adds that "IT CLAIMS THE GREATEST ATTENTION OF BOTH THE PHILOSOPHER AND THEOLOGIAN."

Crown 4to, wrapper, 1s.

JOURNAL OF THE WAGNER SOCIETY.

The Meister.

EDITED BY W. ASHTON ELLIS.

Contains translations from the literary works of Richard Wagner; extracts from letters that have passed between the Poet-Composer and other men who have left their mark upon the art life of the day; original articles and essays explanatory of the inner meaning of Wagner's dramas; articles upon kindred topics of æsthetics, metaphysics, or social questions—in this category, reference to the works of Liszt and Schopenhauer will naturally take a prominent position; notes upon the course of events in Europe and America bearing upon Wagner's dramas, &c., &c.

Third Edition, revised and enlarged.
Crown 8vo, etched Frontispiece and Woodcuts, pp. 324, Cloth gilt, 7s. 6d.

Magic, White and Black;

Or, The Science of Finite and Infinite Life.

CONTAINING PRACTICAL HINTS FOR STUDENTS OF OCCULTISM.

BY FRANZ HARTMANN, M.D.

CONTENTS:—The Ideal—The Real and the Unreal—Form—Life—Harmony—Illusion—Consciousness—Unconsciousness—Transformations—Creation—Light, &c.

The *Saturday Review* says:—"In its closely-printed pages students of occultism will find hints, 'practical' and otherwise, likely to be of great service to them in the pursuit of their studies and researches. . . . A book which may properly have the title of *Magic*, for if the readers succeed in practically following its teaching, they will be able to perform the greatest of all magical feats, the spiritual regeneration of Man. Dr Hartmann's book has also gone into a third edition, and has developed from an insignificant pamphlet, 'written originally for the purpose of demonstrating to a few inexperienced inquirers that the study of the occult side of nature was not identical with the vile practices of sorcery,' into a compendious volume, comprising, we are willing to believe, THE ENTIRE PHILOSOPHIC SYSTEM OF OCCULTISM. There are abundant evidences that the science of theosophy has made vast strides in public estimation of late years, and that those desirous of experimenting in this particular, and in many respects fascinating, branch of ethics, have leaders whose teaching they can follow with satisfaction to themselves."

The *Scotsman* says:—"Any one who studies the work so as to be able to understand it, may become as familiar with the hidden mysteries of nature as any occult philosopher ever was."

*4to, pp. 37, Cloth extra, 3s. 6d.   The woodcuts coloured by hand, 5s.
Issue limited to 400 copies plain and 60 coloured.*

The Dance of Death,

In Painting and in Print.

By T. Tyndall Wildridge.

With Woodcuts.

Probably few subjects have excited more conjecture or given rise to more mistakes than the "Dance of Death." The earliest painting of the Dance is said to be that at Basel in 1431. The first printed edition was published about 1485. The blocks illustrating Mr Wildridge's work are a series found in a northern printing office many years ago. They seem to be of considerable age, and are somewhat close copies of Holbein's designs so far as they go, but in which of the hundred editions they originally appeared has not to the present been ascertained.

Fcap. 8vo, pp. 40, Cloth limp, 1s. 6d.

Light on the Path.

A Treatise written for the Personal Use of Those who are Ignorant of the Eastern Wisdom, and who Desire to Enter within its Influence.

Written down by M. C.

New Edition, with Notes by the Author.

"So far as we can gather from the mystic language in which it is couched, 'Light on the Path' is intended to guide the footsteps of those who have discarded the forms of religion while retaining the moral principle to its fullest extent. It is in harmony with much that was said by Socrates and Plato, although the author does not use the phraseology of those philosophers, but rather the language of Buddhism, easily understood by esoteric Buddhists, but difficult to grasp by those without the pale. 'Light on the Path' may, we think, be said to be THE ONLY ATTEMPT IN THIS LANGUAGE AND IN THIS CENTURY TO PUT PRACTICAL OCCULTISM INTO WORDS; and it may be added, by way of further explanation, that the character of Gautama Buddha, as shown in Sir Edwin Arnolds' 'Light of Asia,' is the perfect type of the being who has reached the threshold of Divinity by this road. That it has reached a third edition speaks favourably for this *multum in parvo* of the science of occultism; and 'M. C.' may be expected to gather fresh laurels in future."—*Saturday Review.*

32mo, pp. 60, Cloth gilt, 1s. 6d.; with pack of 78 Tarot Cards, 5s.

FORTUNE TELLING CARDS.

The Tarot;

Its Occult Signification, Use in Fortune Telling, and Method of Play, &c.

By S. L. MACGREGOR MATHERS.

"The designs of the twenty-one trump cards are extremely singular; in order to give some idea of the manner in which Mr Mather uses them in fortune-telling it is necessary to mention them in detail, together with the general signification which he attaches to each of them. The would-be cartomancer may then draw his own particular conclusions, and he will find considerable latitude for framing them in accordance with his predilections. It should further be mentioned that each of the cards when reversed conveys a meaning the contrary of its primary signification. No. 1 is the Bateleur or Juggler. The Juggler symbolizes Will. 2. The High Priestess, or female Pope, represents Science, Wisdom, or Knowledge. 3. The Empress, is the symbol of Action or Initiative. 4. The Emperor, represents Realization or Development. 5. The Heirophant or Pope, is the symbol of Mercy and Beneficence. 6. The Lovers, signify Wise Disposition and Trials surmounted. 7. The Chariot, represents Triumph, Victory over Obstacles. 8. Themis or Justice, symbolizes Equilibrium and Justice. 9. The Hermit, denotes Prudence. 10. The Wheel of Fortune, represents Fortune, good or bad. 11. Fortitude, symbolizes Power or Might. 12. The Hanged Man—a man suspended head downwards by one leg—means Devotion, Self-Sacrifice. 13. Death, signifies Transformation or Change, 14. Temperance, typifies Combination. 15. The Devil, is the image of Fate or Fatality. 16. The Lightning-struck Tower, called also Maison-Dieu, shows Ruin, Disruption. 17. The Star, is the emblem of Hope. 18. The Moon, symbolizes Twilight, Deception and Error. 19. The Sun, signifies Earthly Happiness. 20. The Last Judgment, means Renewal, Determination of a matter. 21. The Universe, represents Completion and Reward. 0. The Foolish Man, signifies Expiating or Wavering. Separate meanings, with their respective converses, are also attached to each of the other cards in the pack, so that when they have been dealt out and arranged in any of the combinations recommended by the author for purposes of divination, THE INQUIRER HAS ONLY TO USE THIS LITTLE VOLUME AS A DICTIONARY IN ORDER TO READ HIS FATE."—*Saturday Review.*

Crown 8vo, pp. iv. and 256, Cloth (Cheap Edition), 3s. 6d.

A Professor of Alchemy
(*DENIS ZACHAIRE*).

By PERCY ROSS,
AUTHOR OF "A COMEDY WITHOUT LAUGHTER."

"A clever story. . . . The hero is an alchemist who actually succeeds in manufacturing pure gold."—*Court Journal.*

"Shadowy and dream-like."—*Athenæum.*

"An interesting and pathetic picture."—*Literary World.*

"The story is utterly tragical, and is powerfully told."—*Westminster Review.*

"A vivid picture of those bad old times."—*Knowledge.*

"SURE OF A SPECIAL CIRCLE OF READERS WITH CONGENIAL TASTES."—*Graphic.*

"This is a story of love—of deep, undying, refining love—not without suggestions of Faust. The figure of Berengaria, his wife, is a noble and touching one, and her purity and sweetness stand out in beautiful relief from the gloom of the alchemist's laboratory and the horrors of the terrible Inquisition into whose hands she falls. The romance of the crucible, however, is not all permeated by sulphurous vapours and tinged with tartarean smoke. There is often a highly dramatic element."—*Glasgow Herald.*

Fcap. 8vo, pp. 56, Cloth limp, 1s.

The Shakespeare Classical Dictionary;
Or, Mythological Allusions in the Plays of Shakespeare Explained.

FOR THE USE OF SCHOOLS AND SHAKESPEARE READING SOCIETIES.

By H. M. SELBY.

"A handy little work of reference for readers and students of Shakespeare."—*School Board Chronicle.*

"The book presents a great deal of information in a very small compass."—*School Newspaper.*

"Will be found extremely useful by non-classical students of Shakespeare, . . . and even to the classical student it will convey much useful information."—*Educational Times.*

"Will be greatly appreciated in the class-room."—*Glasgow Herald.*

"Carefully compiled from more authoritative books of reference."—*Scotsman.*

"The unlearned reader is thus enabled to increase very greatly his enjoyment of Shakespeare."—*Literary World.*

"WE HAVE TESTED THE BOOK by looking for several of the obscurest mythological names mentioned by Shakespeare; in each case we found the name inserted and followed by a satisfactory explanation."—*The Schoolmaster.*

Demy 8vo, pp. iv. and 299, Cloth gilt, 10s. 6d.

Serpent Worship,

And other Essays, with a Chapter on Totemism.

BY C. STANILAND WAKE.

CONTENTS:—Rivers of Life—Phallism in Ancient Religions—Origin of Serpent Worship—The Adamites—The Descendants of Cain—Sacred Prostitution—Marriage among Primitive Peoples—Marriage by Capture—Development of the "Family"—The Social Position of Woman as affected by "Civilization"—Spiritism and Modern Spiritualism—Totems and Totemism—Man and the Ape.

"The most important of the thirteen essays discusses the origin of Serpent Worship. Like other papers which accompany it, it discusses its subject from a wide knowledge of the literature of early religions and the allied themes of anthropology and primitive marriage. . . . The remaining essays are WRITTEN WITH MUCH LEARNING AND IN A CAREFUL SPIRIT OF INQUIRY, happily free from the crude mysticism with which the discussion of these subjects has often been mixed up. They may be recommended to the attention of all interested in anthropology and the history of religion as interesting labours in this field of research and speculation."—*Scotsman*, October 31.

"So obscure and complex are these subjects that any contribution, however slight, to their elucidation, may be welcomed. Mr Wake's criticism of the systems of others is frequently acute. . . . Mr Wake is opposed to those who hold that kinship through females and the matriarchate preceded paternal kinship and the patriarchal family, and who connect the phenomena of exogamy and of totemism with the matriarchal stage of society, and with belief in a definite kinship of man with the remainder of the sensible universe. He looks upon female kinship as having existed concurrently with a quasi-patriarchal system."—*Athenæum.*

"Able, and REMARKABLY INTERESTING."—*Glasgow Herald.*

Crown 8vo, pp. viii. *and* 632, *Cloth gilt,* 10s. 6d.

In Praise of Ale;

Or, Songs, Ballads, Epigrams, and Anecdotes relating to Beer, Malt, and Hops.

WITH SOME CURIOUS PARTICULARS CONCERNING ALE-WIVES AND BREWERS, DRINKING-CLUBS AND CUSTOMS.

COLLECTED AND ARRANGED BY W. T. MARCHANT.

CONTENTS :—Introductory—History—Carols and Wassail Songs—Church Ales and Observances—Whitsun Ales—Political—Harvest Songs—General Songs—Barley and Malt—Hops—Scotch Ale Songs—Local and Dialect Songs—Trade Songs—Oxford Songs—Ale Wives—Brewers—Drinking Clubs and Customs—Royal and Noble Drinkers—Black Beer—Drinking Vessels—Warm Ale—Facts, Scraps, and Ana.

"Mr Marchant has collected a vast amount of odd, amusing, and (to him that hath the sentiment of beer) suggestive and interesting matter. His volume (we refuse to call it a book) IS A VOLUME TO HAVE. If only as a manual of quotations, if only as a collection of songs, IT IS A VOLUME TO HAVE. We confess to having read in it, for the first time in our lives, the right and authentic text of 'A Cobbler there was' and 'Why, Soldiers, why;' and to have remarked, as regards the first, that our ancestors were very easily amused, and, as regards the second, that it has a curious *air de famille* with the triolet. These are very far from being Mr Marchant's only finds; but that is all the more reason why we should linger upon them."—*Saturday Review.*

"A kind of scrap-book, crowded with prose and verse which is ALWAYS CURIOUS AND VERY OFTEN ENTERTAINING, and it may be read at random—beginning at the end, or in the middle, or at any page you like, and reading either back or forwards—almost as easily as the 'Varieties' column in a popular weekly print."—*Saturday Review.*

"While, on the one hand, the book is, as nearly as possible, a complete collection of lyrics written about the national beverage, . . . it abounds, on the other hand, in particulars as to the place which ale has held in the celebration of popular holidays and customs. It discourses of barley-malt and hops, brewers, drinkers, drinking clubs, drinking vessels, and the like; and, in fact, approaches the subject from all sides, bringing together, in the space of 600 pages, A HOST OF CURIOUS AND AMUSING DETAILS."—*Globe*, April 9.

"Mr Marchant is a staunch believer in the merits of good ale. In the course of his reading he has selected the materials for a Bacchanalian anthology which MAY ALWAYS BE READ WITH AMUSEMENT AND PLEASURE. His materials he has set in a framework of gossiping dissertation. Much curious information is supplied in the various chapters on carols and wassail songs, church ales and observances, Whitsun ales, harvest songs, drinking clubs and customs, and other similar matters. At snug country inns at which the traveller may be called upon to stop there should be, in case of a rainy hour in the day, or an empty smoke-room at night, a copy of a book which sings so loudly the praises of mine host and his wares."—*Notes and Queries.*

"The memory of John Barleycorn is in no danger of passing away for lack of a devoted prophet. The many songs, poems, and pieces of prose written *In Praise of Ale* form a fine garden for the anthologist to choose a bouquet from. . . . It is plainly AN ORIGINAL COLLECTION, made with diligence and good taste in selection. . . . Mr Marchant's anthology may be recommended to the curious as an interesting and carefully compiled collection of poetical and satirical pieces about beer in all its brews."—*Scotsman.*

"The author has gone to ancient and modern sources for his facts, and has not contented himself with merely recording them, but has woven them into a readable history with much skill and wit."—*American Bookseller.*

"Although its chief aim is to be amusing, it is sometimes instructive as well. . . . His stories may at times be a little long, but they are never broad."—*Glasgow Herald.*

"What teetotallers would call A TIPPLER'S TEXT-BOOK . . . a collection of songs and ballads, epigrams and anecdotes, which may be called *unique.*"—*Pall Mall Gazette.*

"Beer, however, in conjunction with mighty roast beef, according to Mr Marchant, has made England what it is, and accordingly he writes his book to show how the English have ever loved good ale, and how much better that is for them than cheap and necessarily inferior spirits or doctored wines. Be that as it may, we have here a collection of occasional verse—satires, epigrams, humorous narratives, trivial ditties, and ballads—VALUABLE AS ILLUSTRATIONS OF MANNERS."—*Literary World.*

Demy 8vo, Cloth, red edges, 7s. 6d.

The Theological and Philosophical Works
OF
Hermes Trismegistus,
CHRISTIAN NEOPLATONIST.

TRANSLATED FROM THE ORIGINAL GREEK, WITH PREFACE, NOTES, AND INDICES.

By JOHN DAVID CHAMBERS, M.A., F.S.A.,
OF ORIEL COLLEGE, OXFORD, RECORDER OF NEW SARUM.

OPINION OF THE AUTHORS OF "THE PERFECT WAY."

"The book is most scholarly and learned, and of great value for its collation of the Bible, Plato, and other Scriptures with the Hermetic, showing one system of thought as pervading them all. He comes to the conclusion —which we also entertain—that the so-called Hermetic books, while representing, in part, ancient Egyptian doctrine, belong to an early Christian— or, perhaps, slightly præ-Christian—period, and are intended to show the identity of the outgoing and incoming systems, and bridge over the gap between them, if any. He omits the *Virgin of the World*, as belonging to some other school, and also the *Asclepius, or Treatise on Initiation*, so that the book does not supersede that which we translated and edited. The author, or rather editor, is not an occultist, but, barring this element, his work is a great addition to Hermetic literature."

Wrapper, price 1s.

Journal of the Bacon Society.

Published Periodically.

Vol. I. (Parts i. to vi.), pp. x. *and* 278, *8vo, cloth,* 6s. 6d.

The main objects for which this Society has been established are :—(*a*) To study the works of Francis Bacon, as Philosopher, Lawyer, Statesman, **and** Poet, also his character, genius, and life, his influence on his own and succeeding times, and the tendencies and results of his writings ; (*b*) **To** investigate Bacon's supposed authorship of certain works unacknowledged by him, including the Shakespearian dramas and poems.

Small 8vo, White Cloth, 4s. 6d.

Through the Gates of Gold:
A Fragment of Thought.

By MABEL COLLINS.

Contents :—The Search for Pleasure—The Mystery of the Threshold—The Initial Effort—The Meaning of Pain—The Secret of Strength.

Crown 8vo, pp. xii. *and* 666, *Cloth,* 10s. 6d.

Myths, Scenes, and Worthies of Somerset.

By Mrs E. BOGER.

Contents :—Bladud, King of Britain ; or, The Legend of Bath—Joseph of Arimathea and the Legend of Glastonbury—Watchet, The Legend of St Decuman—Porlock and St Dubritius—King Arthur in Somerset—St Keyna the Virgin, of Keynsham—Gildas Badonicus, called Gildas the Wise, also Gildas the Querulous—St Brithwald, Archbishop of Canterbury—King Ina in Somerset, Ina and Aldhelm—St Congar and Congresbury—Hun, the Leader of the Sumorsætas, at the Battle of Ellandune—King Alfred in Somerset, and the Legend of St Neot—St Athelm, Archbishop of Canterbury—Wulfhelm, Archbishop of Canterbury—The Landing of the Danes at Watchet—The Times of St Dunstan; His Life and Legends—Muchelney Abbey—Ethelgar, Archbishop of Canterbury—Sigeric or Siricius, Archbishop of Canterbury—Elfeah, Elphége, or Alphege, Archbishop of Canterbury—Ethelnoth, or Ageinoth, Archbishop of Canterbury—Montacute and the Legend of Waltham Cross—Porlock, and Harold son of Godwin—Glastonbury after the Conquest, Bishop Thurstan—William of Malmesbury, called also "Somersetanus"—The Philosophers of Somerset in the Twelfth and Thirteenth Centuries—The Rose of Cannington ; Joan Clifford, commonly called "Fair Rosamond"—John de Courcy—St Ulric the Recluse, or St Wulfric the Hermit—Sir William de Briwere—Woodspring Priory, and the Murderers of Thomas à Becket—Richard of Ilchester, or Richard Tocklive or More—Halswell House, near Bridgewater—The Legend of the House of Tynte—Witham Priory and St Hugh of Avalon (in Burgundy)—William of Wrotham—Joceline Trotman, of Wells

—Hugh Trotman, of Wells—Roger Bacon—Sir Henry Bracton, Lord Chief Justice in the Reign of Henry III.—William Briwere (Briewere, Bruere, or Brewer)—Dunster Castle, Sir Reginald de Mohun, Lady Mohun—Fulke of Samford—Sir John Hautville and Sir John St Loe—Sir Simon de Montacute—The Evil Wedding, Chew Magna and Stanton Drew—Robert Burnel—Somerton, King John of France—Stoke-under-Ham, Sir Matthew Gournay—Bristol (St Mary Redcliffe), The Canyges; Chatterton—Thomas de Beckyngton—The Legend of Sir Richard Whittington—The Legend of the Abbot of Muchelney—Sebastian Cabot—Taunton and its Story—Giles Lord Daubeney and the Cornish Rebellion, King Ina's Palace and South Petherton—John Hooper, The Marian Persecution—The Paulets, Pawlets, or Pouletts, of Hinton St George—Richard Edwardes—Lord Chief Justice Popham—The Last Days of Glastonbury—William Barlow and the Times of Edward VI.—Robert Parsons, or Persons—Henry Cuff—Sir John Harrington—The Wadhams, Wadham College, Oxford; Ilminster, Merrifield, Ilton—Samuel Daniel—Dr John Bull—Thomas Coryate, of Odcombe, in Somerset—John Pym—Sir Amias Preston—Admiral Blake—William Prynne—Sir Ralph, Lord Hopton—Ralph Cudworth—On Witches, Mrs Leakey, of Mynehead, Somerset—John Locke—Thomas Ken, D.D., sometime Bishop of Bath and Wells—Trent House, Charles II. and Colonel Wyndham—The Duke of Monmouth in Somerset—Prince George of Denmark and John Duddleston of Bristol—Beau Nash, with some Account of the Early History of the City of Bath—Wokey or Ockey Hole, near Wells—Captain St Loe—The State of the Church in the Eighteenth Century, Mrs Hannah and Mrs Patty More and Cheddar—Dr Thomas Young—Edward Hawkins, Provost of Oriel and Canon of Rochester—Charles Fuge Lowder—A Tale of Watchet, The Death of Jane Capes—Captain John Hanning Speke—Cheddar Cheese, West Pennard's Wedding Present to the Queen, 1839—In Memoriam, 1811-1833.

"Mrs Boger is to be praised for her enthusiasm and zeal. She is of Somerset, and she naturally thinks it the wonder of England, if not of the world."—*Literary World.*

"Every addition to the local collections of the myths and legends of our country districts is to be welcomed when it is as carefully made as Mrs Boger's laboriously compiled work, which TEEMS WITH QUAINT STORIES, SOME OF WHICH ARE EVEN BEAUTIFUL."—*Westminster Review.*

"This is the kind of book, we imagine, in which Thomas Fuller would have expatiated with delight. Less topographical than his 'Worthies,' it does what that delectable book did not profess to do; it gives not only an account of the illustrious natives, but the legends, traditions, historical episodes, and general *memorabilia* which pertain to one famous county. Mrs Boger's book ranges from Bladud, King of Britain, B.C. 900, to Arthur Hallum, who died in 1833."—*Notes and Queries.*

"Mrs Boger writes with such ability and enthusiasm. The work is one which will have an influence in limits far wider than the borders of Somerset, for FEW CAN READ IT WITHOUT PLEASURE, AND NONE WITHOUT PROFIT. . . . To read her book carefully is to master the hagiology of the county."—*Morning Post.*

GEORGE REDWAY'S

Classified Catalogue of Books,

RELATING TO OCCULT PHILOSOPHY AND ARCHÆOLOGY; EMBRACING COLLECTIONS OF WORKS ON ASTROLOGY, MESMERISM, ALCHEMY, THEOSOPHY, AND MYSTICISM; ANCIENT RELIGIONS AND MYTHOLOGY; ORIENTAL ANTIQUITIES; FREEMASONRY AND SECRET SOCIETIES; WESTERN PHILOSOPHY AND SCIENCE.

"It is certain that one branch at least of historical enquiry—that which deals with the origin and development of religious belief throughout the world—is attracting to itself an increasing degree of attention and interest."—*Quarterly Review, July,* 1886.

The Literature of Occultism and Archæology:

BEING A CATALOGUE OF BOOKS ON SALE RELATING TO

Ancient Worships, Astrology, Alchemy, Animal Magnetism, Anthropology, Arabic, Assassins, Antiquities, Ancient History, Behmen and the Mystics, Buddhism, Clairvoyance, Cabeiri, China, Coins, Druids, Dreams and Visions, Divination, Divining Rod, Demonology, Ethnology, Egypt, Fascination, Flagellants, Freemasonry, Folk Lore, Gnostics, Gems, Ghosts, Hindus, Hieroglyphics and Secret Writing, Herbals, Hermetic, India and the Hindus, Kabbala, Koran, Miracles, Mirabilaries, Magic and Magicians, Mysteries, Mithraic Worship, Mesmerism, Mythology, Metaphysics, Mysticism, Neo-platonism, Orientalia, Obelisks, Oracles, Occult Sciences, Phallic Worship, Philology, Persian, Parsees, Philosophy, Physiognomy, Palmistry and Handwriting, Phrenology, Psychoneurology, Psychometry, Prophets, Rosicrucians, Round Towers, Rabbinical, Spiritualism, Skeptics, Jesuits, Christians and Quakers, Sibylls, Symbolism, Serpent Worship, Secret Societies, Somnambulism, Travels, Tombs, Theosophical, Theology and Criticism, Witchcraft.

"Books on witchcraft, magic, and kindred subjects realize high prices, and a few years hence will be difficult to procure at all, unless, indeed, Mr Redway or some other astute purchaser cares to duplicate his stock while there is time, and keep it under lock and key, for the benefit of the next generation."—*The Athenæum, Feb. 2, 1889.*

List of Books

Chiefly from the Library of the late Frederick Hockley, Esq.,

CONSISTING OF IMPORTANT WORKS RELATING TO THE OCCULT SCIENCES, BOTH IN PRINT AND MANUSCRIPT;

NOW ON SALE AT THE PRICES AFFIXED, BY

GEORGE REDWAY, YORK STREET, COVENT GARDEN, LONDON.

"The study of occultism is not without its charms; and, when an author has anything to say about magic and magicians, about alchemy or astrology, or any other black art, properly so called, he is justified in describing his book as a contribution to the literature of occultism. But the ravings of 'illuminated' persons who have gone mad upon a diet of tetragrams, pentagrams, and pantacles soon pall, and the student turns joyously to the folios of the olden gropers after the Philosopher's Stone. There he finds a treasure of delightful literature, in which amusement is artfully blended with instruction, and where moral maxims are scattered about the pages which teach you how to subject your enemies to a horrible death. The old magicians in their books are equal to any emergency. They will tell you how to raise the devil, and compel him to enrich you with hidden treasures; how to bring the reluctant fair to your arms; how to cast your own nativity; or, if you trouble about none of these things, and incline to lighter sports, they will give you a recipe for charming fish out of the water, or enable you to dream that you are in whatever you may deem to be the right paradise. With speculations about the why and the wherefore of things they will not trouble you. They prefer to dilate upon the wonders of black magic, and to gloat over the one hundred thousand pounds' weight of fine gold which a friend of Raymond Lully's made by alchemical means. These musty tomes, full of significant circles and magic triangles, of red dragons and black hens, embellished with portraits of the demoniacal hierarchy and drawings of the essential implements for evoking spirits, have a pleasant flavour of romance. The quaint Latinity and the odd jumble of tongues in which the conjurations are written are as fine in their way as anything that ever was printed in a folio. But it is needful to beware of the endless volumes of modern ravings about the so-called occult; for that way madness lies."—*Saturday Review, April 23, 1887.*

Crown 8vo, pp. 375, Cloth, 7s. 6d.

Theosophy, Religion, and Occult Science.

By HENRY S. OLCOTT,
PRESIDENT OF THE THEOSOPHICAL SOCIETY.

WITH GLOSSARY OF EASTERN WORDS.

CONTENTS:—Theosophy or Materialism—Which?—The Theosophical Society and its Aims—The Common Foundation of all Religions—Thesophy: the Scientific Basis of Religion—Theosophy: its Friends and Enemies—The Occult Sciences—Spiritualism and Theosophy—India: Past, Present, and Future—The Civilisation that India needs—The Spirit of the Zoroastrian Religion—the Life of Buddha and its Lessons, &c.

The *Manchester Examiner* describes these lectures as "RICH IN INTEREST AND SUGGESTIVENESS," and says that "the theosophy expounded in this volume is at once a theology, a metaphysic, and a sociology," and concludes a lengthy notice by stating that "Colonel Olcott's volume deserves, and will repay, the study of all readers for whom the byways of speculation have an irresistible charm."

Demy 8vo, pp. xii. and 324, Cloth, 10s. 6d.

Incidents in the Life of Madame Blavatsky.

COMPILED FROM INFORMATION SUPPLIED BY HER RELATIVES AND FRIENDS,

AND EDITED BY A. P. SINNETT.

WITH A PORTRAIT REPRODUCED FROM AN ORIGINAL PAINTING BY HERMANN SCHMIECHEN.

CONTENTS:—Childhood—Marriage and Travel—At Home in Russia, 1858—Mme. de Jelihowsky's Narrative—From Apprenticeship to Duty—Residence in America—Established in India—A Visit to Europe, &c.

Truth says:—"For any credulous friend who revels in such stories I can recommend 'Incidents in the Life of Madame Blavatsky.' I READ EVERY LINE OF THE BOOK WITH MUCH INTEREST."

Theosophists will find both edification and interest in the book.

Post 8vo, pp. viii. *and* 350, *Cloth gilt,* 7s. 6d.

The Blood Covenant, a Primitive Rite,

And its Bearings on Scripture.

By H. CLAY TRUMBULL, D.D.

CONTENTS:—*The Primitive Rite Itself.*—(1) Sources of Bible Study—(2) An Ancient Semitic Rite—(3) The Primitive Rite in Africa—(4) Traces of the Rite in Europe—(5) World-wide Sweep of the Rite,—(6) Light from the Classics—(7) The Bond of the Covenant,—(8) The Rite and its Token in Egypt—(9) Other Gleams of the Rite. *Suggestions and Perversions of the Rite.*—(1) Sacredness of Blood and of the Heart—(2) Vivifying Power of Blood—(3) A new Nature through new Blood—(4) Life from any Blood, and by a Touch—(5) Inspiration through Blood—(6) Inter-communion through Blood—(7) Symbolic Substitutes for Blood—(8) Blood Covenant Involvings. *Indications of the Rite in the Bible.*—(1) Limitations of Inquiry—(2) Primitive Teachings of Blood—(3) The Blood Covenant in Circumcision—(4) The Blood Covenant Tested—(5) The Blood Covenant and its Tokens in the Passover—(6) The Blood Covenant at Sinai—(7) The Blood Covenant in the Mosaic Ritual—(8) The Primitive Rite Illustrated—(9) The Blood Covenant in the Gospels—(10) The Blood Covenant applied. Importance of this Rite strangely undervalued—Life in the Blood, in the Heart, in the Liver—Transmigration of Souls—The Blood-rite in Burmah—Blood-stained Tree of the Covenant—Blood-drinking—Covenant Cutting—Blood-bathing—Blood-ransoming—The Covenant-reminder —Hints of Blood Union—Topical Index—Scriptural Index.

"An admirable study of a primitive belief and custom—one of the utmost importance in considering the growth of civilisation. . . . In the details of the work will be found much to attract the attention of the curious. Its fundamental and essential value, however, is for the student of religions; and all such will be grateful to Dr Trumbull for THIS SOLID, INSTRUCTIVE, AND ENLIGHTENING WORK."—*Scotsman.*

Square 16mo, *Cloth, gilt edges,* 5s.

The Art of Judging the Character of Individuals

FROM

their Handwriting and Style.

WITH 35 PLATES, CONTAINING 120 SPECIMENS OF THE HANDWRITING OF VARIOUS CHARACTERS.

EDITED BY EDWARD LUMLEY.

CONTENTS, AND LIST OF PLATES.—(1) Art of Judging the Character by the Handwriting, now first translated from the French: *a.* Introduction; *b.* Character of Men from the Handwriting; *c.* Art of Judging Men by their Style (*Plates* 1 to 22)—(2) Account of alleged Art of Reading the Character of Individuals in their Handwriting, by Dr W. Seller (*Plates* 23, 24, 25)—(3) On Characteristic Signatures, by Stephen Collet, A.M. (*Thomas Byerley*) (*Plates* 26 to 32)—(4) Autographs, by Isaac D'Israeli—(5) Hints as to Autographs, by Sir John Sinclair—(6) Characters in Writing, by Vigneul Marville (*Dom Noel Dargonne*)—(7) The Autograph a Test of Character, by Edgar A. Poe (*Plates* 33, 34)—(8) Of Design, Colouring, and Writing, by the Rev. J. Casper Lavater (*Plate* 35).

Post 8vo, pp. xiii. *and* 220, *Cloth,* 10s. 6d.

The Life

OF

Philippus Theophrastus, Bombast of Hohenheim,

KNOWN BY THE NAME OF

Paracelsus.

AND THE SUBSTANCE OF HIS TEACHINGS CONCERNING COSMOLOGY, ANTHROPOLOGY, PNEUMATOLOGY, MAGIC AND SORCERY, MEDICINE, ALCHEMY AND ASTROLOGY, PHILOSOPHY AND THEOSOPHY.

EXTRACTED AND TRANSLATED FROM HIS RARE AND EXTENSIVE WORKS, AND FROM SOME UNPUBLISHED MANUSCRIPTS,

BY FRANZ HARTMANN, M.D.

CONTENTS:—The Life of Paracelsus—Explanation of Terms—Cosmology—Anthropology—Pneumatology—Magic and Sorcery — Medicine—Alchemy and Astrology—Philosophy and Theosophy—Appendix.

St James's Gazette describes this as "a book which will have some permanent value to the student of the occult," and says that "STUDENTS SHOULD BE GRATEFUL FOR THIS BOOK, despite its setting of Theosophical nonsense."

Crown 8vo, pp. x. *and* 124, *Parchment,* 6s.

The Raven.

BY EDGAR ALLAN POE.

WITH LITERARY AND HISTORICAL COMMENTARY BY JOHN H. INGRAM.

CONTENTS:—Genesis—The Raven, with Variorum Readings—History—Isadore—Translations : French — German — Hungarian — Latin—Fabrications—Parodies—Bibliography—Index.

"An interesting monograph on Poe's famous poem."—*Spectator.*

"THERE IS NO MORE RELIABLE AUTHORITY ON THE SUBJECT THAN MR JOHN H. INGRAM. Much curious information is collected in his essay. The volume is well printed and tastefully bound in spotless vellum."—*Publishers' Circular.*

Crown 8vo, pp. xxviii. and 184, Cloth, 5s.

The History of Tithes,

From Abraham to Queen Victoria.

By HENRY W. CLARKE.

Contents:—The History of Tithes before the Christian Era—From the Christian Era to A.D. 400—From A.D. 400 to A.D. 787—From A.D. 787 to A.D. 1000—From A.D. 1000 to A.D. 1215—From A.D. 1215 to the Dissolution of Monasteries—Monasteries—Infeudations—Exemption from Paying Tithes—The Dissolution of Monasteries—The Commutation Act of 1836, 6 and 7 Will. IV., c. 71—Tithes in the City and Liberties of London—Redemption of Tithe Rent Charge—Some Remarks on "A Defence of the Church of England against Disestablishment," by the Earl of Selborne.

"An impartial and valuable array of facts and figures, which should be read by all who are interested in the solution of the tithe problem."—*Athenæum.*

"THE BEST BOOK OF MODERATE SIZE YET PUBLISHED for the purpose of enabling an ordinary reader to thoroughly understand the origin and history of this ancient impost."—*Literary World.*

Crown 8vo., pp. viii. and 184, Cloth, 2s. 6d.

Burma as it was, as it is, and as it will be.

By JAMES GEORGE SCOTT.

(*Shway Yoe.*)

Contents:—I. The History—Burma according to Native Theories—Origin of the Burmese—Early History—First appearance of Europeans in Burma—Worrying our Representatives—War with Burma—The Inevitable End. II. The Country—Lower Burma—Upper Burma—The Irrawaddy to Mandalay—Mandalay—The Irrawaddy above Mandalay. III. The People—Burmese Kings—Burmese Officials—The Hloat-daw—The Officers of the Household—Method of Appointment and Payment—The People—Their Faults—Excellence as Buddhists—Doctrine of Good Works—Superstitions—Lucky and Unlucky Days—The most Sociable of Men—Freedom of the Women—A Nation of Smokers—Contented with British Rule—Ascendency of the Chinaman Trade—Hill-tribes—Their Religion—Hope for the Nomads—The Kachyens.

The *Saturday Review* says:—"Before going to help to govern them, Mr Scott has once more written on the Burmese . . . Mr Scott claims to have covered the whole ground, and as there is nobody competent to criticise him except himself, we shall not presume to say how far he has succeeded. What, however, may be asserted with absolute confidence is, that he has written A BRIGHT, READABLE, AND USEFUL BOOK."

LARGE PAPER EDITION, Royal 8vo, pp. xvi. and 60, 7s. 6d.

An Essay on the Genius of George Cruikshank.

BY WILLIAM MAKEPEACE THACKERAY.

Reprinted Verbatim from " The Westminster Review."

EDITED WITH A PREFATORY NOTE ON THACKERAY AS AN ARTIST AND ART CRITIC, BY W. E. CHURCH.

WITH UPWARDS OF FORTY ILLUSTRATIONS, INCLUDING ALL THE ORIGINAL WOODCUTS, AND A NEW PORTRAIT OF CRUIKSHANK ETCHED BY F. W. PAILTHORPE.

As the original copy of the *Westminster* is now excessively rare, this re-issue will no doubt be welcomed by collectors. The new portrait of Cruikshank by F. W. Pailthorpe is a clear firm etching.

Pp. 102, Cloth, 2s. 6d.

Pope Joan
(THE FEMALE POPE);

A Historical Study.

TRANSLATED FROM THE GREEK OF EMMANUEL RHOÏDIS, WITH PREFACE BY

CHARLES HASTINGS COLLETTE.

FRONTISPIECE TAKEN FROM THE ANCIENT MS. NUREMBERG CHRONICLE, PRESERVED AT COLOGNE.

" The subject of Pope Joan will always have its attractions for the lovers of the curiosities of history. Rhoïdis discusses the topic with much learning and ingenuity, and Mr Collette's Introduction is full of information."—*Globe*.

Crown 8vo, pp. 40, printed on hand-made paper, Vellum Gilt, 6s.

The Bibliography of Swinburne;

A BIBLIOGRAPHICAL LIST, ARRANGED IN CHRONOLOGICAL ORDER, OF THE PUBLISHED WRITINGS, IN VERSE AND PROSE, OF ALGERNON CHARLES SWINBURNE (1857-1887).

Only 250 copies printed. The compiler, writing on April 5, 1887, says:—
"Born on April 5, 1837, in the year of Queen Victoria's Accession, of which the whole nation is now celebrating the Jubilee, Algernon Charles Swinburne to-day attains the jubilee or 50th year of his own life, and **may** therefore be claimed **as** an essentially and exclusively Victorian poet."

INDISPENSABLE TO SWINBURNE COLLECTORS.

Demy 8vo, pp. xxiv. and 104, Cloth extra, 7s. 6d.

The Astrologer's Guide
(ANIMA ASTROLOGIÆ);
Or, A Guide for Astrologers.

BEING

THE ONE HUNDRED AND FORTY-SIX CONSIDERATIONS OF THE FAMOUS ASTROLOGER, GUIDO BONATUS, TRANSLATED FROM THE LATIN BY HENRY COLEY,

TOGETHER WITH

THE CHOICEST APHORISMS OF THE SEVEN SEGMENTS OF JEROME CARDAN OF MILAN, EDITED BY WILLIAM LILLY (1675).

NOW FIRST REPUBLISHED FROM A UNIQUE COPY OF THE ORIGINAL EDITION, WITH NOTES AND A PREFACE, BY

WM. C. ELDON SERJEANT,
FELLOW OF THE THEOSOPHICAL SOCIETY.

"Mr Serjeant deserves the thanks of all who are interested in astrology for rescuing this important work from oblivion. . . . The growing interest in mystical science will lead to a revival of astrological study, and ADVANCED STUDENTS WILL FIND THIS BOOK AN INDISPENSABLE ADDITION TO THEIR LIBRARIES. The book is well got up and printed."—*Theosophist.*

16mo, pp. xvi. and 148, Cloth extra, 2s.

Tobacco Talk and Smokers' Gossip.

AN AMUSING MISCELLANY OF FACT AND ANECDOTE RELATING TO THE "GREAT PLANT" IN ALL ITS FORMS AND USES, INCLUDING A SELECTION FROM NICOTIAN LITERATURE.

CONTENTS:—A Tobacco Parliament—Napoleon's First Pipe—A Dutch Poet and Napoleon's Snuff-Box—Frederick the Great as an Ass—Too Small for Two—A Smoking Empress—The Smoking Princesses—An Incident on the G.W.R—Raleigh's Tobacco Box—Bismarck's Last Cigar—Bismarck's Cigar Story—Moltke's Pound of Snuff—Lord Brougham as a Smoker—Mazzini's Sang-froid as a Smoker—Lord Clarendon as a Smoker—Politics and Snuff-Boxes—Penn and Tobacco—Tobacco and the Papacy—The Snuff-Mull in the Scotch Kirk—Whateley as a Snuff-Taker—The First Bishop who Smoked—Pigs and Smokers—Jesuits' Snuff—Kemble Pipes—An Ingenious Smoker—Anecdote of Dean Aldrich—Smoking to the Glory of God—Professor Huxley on Smoking—Blucher's Pipe-Master—Shakespeare and Tobacco—Ben Jonson on Tobacco—Lord Byron on Tobacco—Decamps and Horace Vernet—Milton's Pipe—Anecdote of Sir Isaac Newton—Emerson and Carlyle—Paley and his Pipe—Jules Sandeau on the Cigar—The Pickwick of Fleet Street—The *Obsequio* of Havana—The Social Pipe (*Thackeray*)—Triumph of Tobacco over Sack and Ale—The Smoking Philosopher—Sam Slick on the Virtues of a Pipe—Smoking in 1610—Bulwer-Lytton on Tobacco-Smoking—Professor Sedgwick—St Pierre on the Effect of Tobacco—Ode to Tobacco (*C. S. Calverley*)—Meat and Drink (*Charles Kingsley*)—The Meerschaum (*O. W. Holmes*)—Charles Kingsley at Eversley—Robert Burns's Snuff-Box—Robinson Crusoe's Tobacco—Guizot—Victor Hugo—Buckle as a Smoker—Carlyle on Tobacco—A Poet's Pipe (*Baudelaire*)—A Pipe of Tobacco—The Headsman's Snuff-box—The Pipe and Snuff-box (*Cowper*)—Anecdote of Charles Lamb—Gibbon as a Snuff-Taker—Charles Lamb as a Smoker—Farewell to Tobacco (*Chas. Lamb*)—The Power of Smoke (*Thackeray*)—Thackeray as a Smoker—Dickens as a Smoker—Chewing and Spitting in America—Tennyson as a Smoker—A Smoker's Opinion of Venice—Coleridge's First Pipe—Richard Porson—Cruikshank and Tobacco—Mr James Payn—Mr Swinburne on Raleigh—The Anti-Tobacco Party—" This Indian Weed "—Dr Abernethy on Snuff-Taking—Abernethy and a Smoking Patient—Tobacco and the Plague—" The Greatest Tobacco Stopper in all England "—Dr Richardson on Tobacco—Advice to Smokers—Some Strange Smokers—The Etymology of Tobacco—The Snuff called "Irish Blackguard"—A Snuff-Maker's Sign—Mr Sala's Cigar-Shop—Death of the "Yard of Clay"—A Prodigious Smoker—A Professor of Smoking—Tobacco in Time of War—Ages attained by Great Smokers—A Maiden's Wish—" Those Dreadful Cigars "—How to take a Pinch of Snuff—The Tobacco Plant—Fate of an Early Smoker—Adding Insult to Injury—Tom Brown on Smoking—The Snuff-Taker—Tobacco in North America—National Characteristics—Smoking at School—Carlyle on "The Veracities "—Children's Pipes—The Uses of Cigar Ash—An Inveterate Smoker—A Tough Yarn—Some French Smokers—Riddles for Smokers—Cigar Manufacturing in Havana.

" One of the best books of gossip we have met for some time. . . . It is literally crammed full from beginning to end of its 148 pages with well-selected anecdotes, poems, and excerpts from tobacco literature and history."—*Graphic.*

" The smoker should be grateful to the compilers of this pretty little volume. . . . NO SMOKER SHOULD BE WITHOUT IT, and anti-tobacconists have only to turn over its leaves to be converted."—*Pall Mall Gazette.*

" Something to please smokers ; and non-smokers may be interested in tracing the effect of tobacco—the fatal, fragrant herb—on our literature."—*Literary World.*

Demy 8vo, pp. xliii. and 349, with Illustrations, Cloth extra, 10s. 6d.

The Mysteries of Magic;

A Digest of the Writings of Éliphas Lévi.

WITH BIOGRAPHICAL AND CRITICAL ESSAY

BY ARTHUR EDWARD WAITE.

CONTENTS:—INITIATORY EXERCISES AND PREPARATIONS—RELIGIOUS AND PHILOSOPHICAL PROBLEMS AND HYPOTHESES—The Hermetic Axiom, Faith—The True God—The Christ of God—Mysteries of the Logos—The True Religion—The Reason of Prodigies, or the Devil before Science—SCIENTIFIC AND MAGICAL THEOREMS—On Numbers and their Virtues—Theory of Will Power—The Translucid—The great Magic Agent, or the Mysteries of the Astral Light—Magic Equilibrium—The Magic Chain—The great Magic Arcanum—THE DOCTRINE OF SPIRITUAL ESSENCES, OR KABBALISTIC PNEUMATICS; WITH THE MYSTERIES OF EVOCATION, NECROMANCY, AND BLACK MAGIC—Immortality—The Astral Body—Unity and Solidarity of Spirits—The great Arcanum of Death, or Spiritual Transition, Hierarchy, and Classification of Spirits—Fluidic Phantoms and their Mysteries—Elementary Spirits and the Ritual of their Conjuration—Necromancy—Mysteries of the Pentagram and other Pantacles—Magical Ceremonial and Consecration of Talismans—Black Magic and the Secrets of the Witches'—Sabbath—Witchcraft and Spells—The Key of Mesmerism—Modern Spiritualism—THE GREAT PRACTICAL SECRETS OR REALISATIONS OF MAGICAL SCIENCE—The "Magnum Opus"—The Universal Medicine—Renewed Youth—Transformations—Divination—Astrology—The Tarot, the Book of Hermes, or of Koth—Eternal Life, or Profound Peace—EPILOGUE—SUPPLEMENT—The Kabbalah—Thaumaturgical Experiences of Eliphas Lévi—Evocation of Apollonius of Tyana—Ghosts in Paris—The Magician and the Medium—Eliphas Lévi and the Sect of Eugène Vintras—The Magician and the Sorcerer—Secret History of the Assassination of the Archbishop of Paris—NOTES.

"Of the many remarkable men who have gained notoriety by their proficiency, real or imaginary, in the Black Arts, probably none presents a more strange and irreconcileable character than the French magician Alphonse Louis Constant. . . . Better known under the Jewish pseudonym of Éliphas Lévi Zahed, this enthusiastic student of forbidden art made some stir in France, and even in London. . . . HIS WORKS ON MAGIC ARE THOSE OF AN UNDOUBTED GENIUS, and divulge a philosophy beautiful in conception, if totally opposed to common sense principles. . . . There is so great a fund of learning and of attractive reasoning in these writings, that Mr Arthur Edward Waite has published a digest of them for the benefit of English readers. This gentleman has not attempted a literal translation in every case, but has arranged a volume which, while reproducing with sufficient accuracy a great portion of the more interesting works, affords an excellent idea of the scope of the entire literary remains of an enthusiast for whom he entertains a profound admiration. . . . The reader may with profit peruse carefully the learned dissertations penned by M. Constant upon the Hermetic art treated as a religion, a philosophy, and a natural science. . . . In view of the remarkable exhibitions of mesmeric influence and thought reading which have been recently given, it is not improbable that the thoughtful reader may find a clue in the writings of this cultured and amiable magician to the secret of many of the manifestations of witchcraft that formerly struck wonder and terror into the hearts of simple folks. . . ."—*The Morning Post.*

"The present single volume is a digest of half-a-dozen books enumerated by the present author in a 'biographical and critical essay' with which he prefaces his undertaking. These are the *Dogme et Rituel de la Haute Magie*, the *Histoire de la Magie*, the *Clef des Grands Mystères*, the *Sorcier de Mendon*, the *Philosophie Occulte*, and the *Science des Esprits*. To attack the whole series—which, indeed, it might be difficult to obtain now in a complete form—would be a bold undertaking, but Mr Waite has endeavoured to give his readers the essence of the whole six books in a relatively compact compass. . . . THE BOOK BEFORE US IS ENCYCLOPÆDIC IN ITS RANGE, and it would be difficult to find a single volume which is better calculated to supply modern inquiries with a general conception of the scope and purpose of the occult sciences at large. It freely handles, amongst others, the ghastly topics of witchcraft and black magic, but certainly it would be difficult to imagine any reader tempted to enter those pathways of experiment by the picture of their character and purpose that Éliphas Lévi supplies. In this way the intrepid old Kabbalist, though never troubling his readers with sublime exhortations in the interests of virtue, writes under the inspiration of an uncompromising devotion to the loftiest ideals, and all his philosophy 'makes for righteousness.' "—Mr A. P. Sinnett in *Light*.

"We are grateful to Mr Waite for translating the account of how Lévi, in a lone chamber in London, called up the spirit of Apollonius of Tyana. This very creepy composition is written in quite the finest manner of the late Lord Lytton when he was discoursing upon the occult."—*The Saturday Review*.

Demy 18mo, *pp.* vi. *and* 132, *with Woodcuts, Fancy Cloth*, 1s.

John Leech, Artist and Humourist.

A Biographical Sketch.

By FRED. G. KITTON.

New Edition, Revised.

"In the absence of a fuller biography we cordially welcome Mr Kitton's interesting little sketch."—*Notes and Queries*.

"The multitudinous admirers of the famous artist will find this touching monograph well worth careful reading and preservation."—*Daily Chronicle*.

"THE VERY MODEL OF WHAT SUCH A MEMOIR SHOULD BE."—*Graphic*.

4to, with Frontispiece, pp. xxx. *and* 154, *Parchment,* 10s. 6d.

THE HERMETIC WORKS.

The Virgin of the World

OF

Hermes Mercurius Trismegistus.

Now first Rendered into English, with Essay, Introductions, and Notes,

By DR ANNA KINGSFORD AND EDWARD MAITLAND,

AUTHORS OF "THE PERFECT WAY."

Published under the auspices of the Hermetic Society. Essays on "The Hermetic Books," by E. M., and on "The Hermetic System and the Significance of its Present Revival," by A. K. "The Virgin of the World" is followed by "Asclepios on Initiation," the "Definitions of Asclepios," and the "Fragments of Hermes."

"It will be a most interesting study for every occultist to compare the doctrines of the ancient Hermetic philosophy with the teaching of the Vedantic and Buddhist systems of religious thought. THE FAMOUS BOOKS OF HERMES seem to occupy, with reference to the Egyptian religion, the same position which the Upanishads occupy in Aryan religious literature."—*Theosophist,* November, 1885.

Imperial 16*mo, pp.* 16, *wrapper, printed on Whatman's hand-made paper.* 250 *copies only, each numbered.* 5s.

A Word for the Navy.

By ALGERNON CHARLES SWINBURNE.

"Mr Swinburne's new patriotic song, 'A Word for the Navy,' is as fiery in its denunciation of those he believes to be antagonistic to the welfare of the country as was his lyric with which he startled the readers of the *Times* one morning."—*Athenæum.*

The publisher of this poem is also the sole proprietor of the *copyright*; it cannot therefore be included in Mr Swinburne's *collected works.*

4to, pp. 121, *Illustrated with a number of beautiful Symbolical Figures,
Parchment gilt, price* 10s. 6d.

ASTROLOGY THEOLOGIZED.

The Spiritual Hermeneutics of Astrology and Holy Writ.

BEING A TREATISE UPON THE INFLUENCE OF THE STARS ON MAN AND ON THE ART OF RULING THEM BY THE LAW OF GRACE.

(*Reprinted from the original of* 1649.)

WITH A PREFATORY ESSAY ON THE TRUE METHOD OF INTERPRETING HOLY SCRIPTURE,

BY ANNA BONUS KINGSFORD.

ILLUSTRATED WITH ENGRAVINGS ON WOOD.

CONTENTS:—What Astrology is, and what Theology; and how they have reference one to another—Concerning the Subject of Astrology—Of the three parts of Man: Spirit, Soul, and Body, from whence every one is taken, and how one is in the other—Of the Composition of the Microcosm, that is Man, from the Macrocosm, the great World—That all kind of Sciences, Studies, Actions, and Lives, flourishing amongst Men on the Earth and Sea, do testify that all Astrology, that is, Natural Wisdom, with all its Species, is and is to be really found in every Man. And so all things, whatsoever Men act on Earth, are produced, moved, governed, and acted from the Inward Heaven. And what are the Stars which a Wise Man ought to rule. Touching a double Firmament and Star in every Man; and that by the Benefit of Regeneration in the Exercise of the Sabbath, a Man may be transposed from a worse nature into a better—Touching the Distribution of all Astrology into the Seven Governors of the World, and their Operations and Offices, as well in the Macrocosm as in the Microcosm—Touching the Astrology of Saturn, of what kind it is, and how it ought to be Theologized—A Specifical Declaration, how the Astrology of Saturn in Man ought to be and may be Theologized.

The *St James's Gazette* says:—" It is well for Dr Anna Kingsford that she was not born into the sidereal world four hundred years ago. Had that been her sorry fate, she would assuredly have been burned at the stake for her preface to 'Astrology Theologized.' It is a very long preface—more than half the length of the treatise it introduces; IT CONTAINS SOME OF THE FINEST FLOWERS OF THEOSOPHICAL PHILOSOPHY, and of course makes very short work of Christianity."

Crown 8vo, pp. 56, *printed on Whatman's Handmade Paper, Vellum Gilt,* 6s.

Hints to Collectors

Of Original Editions of the Works of Charles Dickens.

By CHARLES PLUMPTRE JOHNSON.

Including Books, Plays, and Portraits, there are 167 items fully described.

"This is a sister volume to the 'Hints to Collectors of First Editions of Thackeray,' which we noticed a month or two ago. As we are unable to detect any slips in his work, we must content ourselves with thanking him for the correctness of his annotations. It is unnecessary to repeat our praise of the elegant *format* of these books."—*Academy.*

Crown 8vo, pp. 48, *printed on Whatman's Handmade Paper, Vellum Gilt,* 6s.

Hints to Collectors

Of Original Editions of the Works of William Makepeace Thackeray.

By CHARLES PLUMPTRE JOHNSON.

". . . . A guide to those who are great admirers of Thackeray, and are collecting first editions of his works. The dainty little volume, bound in parchment and printed on hand-made paper, is very concise and convenient in form; on each page is an exact copy of the title-page of the work mentioned thereon, a collation of pages and illustrations, useful hints on the differences in editions, with other matters INDISPENSABLE TO COLLECTORS. . . . Altogether it represents a large amount of labour and experience."—*Spectator.*

Large Crown 8vo, pp. xxxii. and 324, Cloth extra, Gilt Top, 10s. 6d.

Sea Song and River Rhyme,

From Chaucer to Tennyson.

SELECTED AND EDITED BY

ESTELLE DAVENPORT ADAMS.

WITH A NEW POEM BY ALGERNON CHARLES SWINBURNE.

WITH TWELVE ETCHINGS.

In general, the Songs and Poetical Extracts are limited to those which deal with the Sea and Rivers as natural objects, and are either descriptive or reflective. The Etchings are printed in different colours; the headpieces are also original.

"The book is, on the whole, *one of the best of its kind ever published.*"—*Glasgow Herald.*

"The editor has made the selection with praiseworthy judgment."—*Morning Post.*

"Twelve really exquisite and delicately executed etchings of sea and riverside accompany and complete THIS BEAUTIFUL VOLUME."—*Morning Post.*

"A special anthology, delightful in itself, and possessing the added graces of elegant printing and dainty illustrations."—*Scotsman.*

"The volume is got up in the handsomest style, and includes a dozen etchings of sea and river scenes, some of which are exquisite."—*Literary World.*

Crown 8vo, pp. xl. and 420, Cloth extra, 10s. 6d.

The History of the Forty Vezirs;

Or, The Story of the Forty Morns and Eves.

WRITTEN IN TURKISH BY SHEYKH-ZĀDA;

DONE INTO ENGLISH BY E. J. W. GIBB, M.R.A.S.

The celebrated Turkish romance, translated from a printed but undated text procured a few years ago in Constantinople.

"A delightful addition to the wealth of Oriental stories available to English readers. . . . Mr Gibb has considerately done everything to help the reader to an intelligent appreciation of THIS CHARMING BOOK."—*Saturday Review.*

SIR RICHARD F. BURTON says:—"In my opinion, the version is definite and final. The style is light and pleasant, with the absolutely necessary flavour of quaintness; and the notes, though short and few, are sufficient and satisfactory."

Complete in 12 Vols. £3 nett.

The Antiquarian Magazine and Bibliographer.

Edited by
EDWARD WALFORD, M.A. and G. W. REDWAY, F.R.H.S.

This illustrated periodical, highly esteemed by students of English antiquities, biography, folk-lore, bibliography, numismatics, genealogy, &c., was founded in 1882 by Mr Edward Walford, and completed in 1887 under the editorship of Mr G. W. Redway. ONLY SOME THIRTY COMPLETE SETS REMAIN, and they are offered at a very moderate price.

CONTENTS OF VOLS. XI. AND XII.:—Domesday Book — Frostiana — Some Kentish Proverbs—The Literature of Almanacks—"Madcap Harry" and Sir John Popham—Tom Coryate and his Crudities—Notes on John Wilkes and Boswell's Life of Johnson—The Likeness of Christ—The Life, Times, and Writings of Thomas Fuller—Society in the Elizabethan Age—Chapters from Family Chests—Collection of Parodies—Rarities in the Locker-Lampson Collection—A Day with the late Mr Edward Solly—The Defence of England in the 16th Century—The Ordinary from Mr Thomas Jenyn's Booke of Armes—A Forgotten Cromwellian Tomb—Visitation of the Monasteries in the Reign of Henry the Eighth—The Rosicrucians—The Seillière Library—A Lost Work—Romances of Chivalry—Ancient Legends, Mystic Charms, and Superstitions of Ireland—The Art of the Old English Potter—The Story of the Spanish Armada—Books for a Reference Library—Myth-Land—Sir Bevis of Hampton—Cromwell and the Saddle Letter of Charles I.—Recent Discoveries at Rome—Folk-Lore of British Birds—An old Political Broadside—Notes for Coin Collectors—Higham Priory—By-Ways of Periodical Literature—Memoir of Captain Dalton—A History of the Parish of Mortlake, in the County of Surrey—Historic Towns—Exeter—Traits and Stories of Ye Olde Cheshire Cheese—The Pre-History of the North—The Vision of William concerning Piers the Plowman—The Curiosities of Ale—The Books and Bookmen of Reading—How to trace a Pedigree—The Language of the Law—Words, Idioms, &c., of the Vulgar—The Romans in Cumbria—The Study of Coins—An Un-bowdlerised Boccaccio—The Kabbalah—The House of Aldus—Bookselling in Little Britain—Copper-plates and Woodcuts by the Bewicks—Excavations at Ostia—Sir Sages of Somerset—The Good Queen Bertha—The popular Drama of the Past—Relics of Astrologic Idioms—A Leaf from an Old Account Book—The Romance of a Gibbet—General Pardons—Thorscross or Thurscross(Yorkshire)—The Genesis of "In Memoriam"—The Influence of Italian upon English Literature—The Trade Signs of Essex—The Ancient Cities of the New World—The Legendary History of the Cross—History of Runcorn—The Rosicrucians; their Rites and Mysteries—Old Glasgow Families—The House of Aldus—Merlin, the Prophet of the Celts—A facetious Advertisement—Funeral Garlands—Bookselling on London Bridge—Millom Cumberland—A forgotten Children's Book of Charles Dickens—The Rothschilds; a Trilogy of the Life to come—The Beer of the Bible—Story of the Drama in Exeter—By-Ways of Periodical Literature—Reading Anecdotes—Tennysonian and Thackerayan Rarities—The Origin and History of Change Ringing—More Vulgar Words and Phrases—The popular Drama of the past—Some Poems attributed to Byron—The Marriage of Cupid and Psyche—Sketches of Life in Japan—The first nine years of the Bank of England—The Brunswick Accession—History of the Bassandyne Bible—Peculiar Courts—Vulgar Etymologies—Nuremburg—Metal Pan-making in England—The Pews of the Past—Octocentenary of the Death of William the Conqueror—A Black Magician—The Allegorical Signification of the Tinctures in Heraldry—The Purpose of the Ages—The Sieges of Pontefract Castle—A Life of John Colet—The History of Sport in Cheshire—Tom Coryat and his Crudities—The Tarot: an Antique Method of Divination—Law French—The Pews of the Past—Shropshire Folk-Lore—The Printed Book—St Mary Overies Priory Church, Southwark—Some curious passages from Baker's Chronicle—The resting-place of Cromwell—A Library of Rarities—Europe in the reign of James the Sixth—Myths, Scenes, and Worthies of Somerset—Herefordshire Words and Phrases—Chronicles of an Old Inn—Epitaphs—The Gnostics and their Remains—Collectanea—Meetings of Learned Societies—News and Notes—Obituary Memoirs—Correspondence—Vos Valete et Plaudite.

Large Demy 8vo, pp. xx. and 268, Cloth, 10*s.* 6*d.*

Sultan Stork;

And other Stories and Sketches.

BY WILLIAM MAKEPEACE THACKERAY.
(1829-1844.)

NOW FIRST COLLECTED.

TO WHICH IS ADDED THE BIBLIOGRAPHY OF THACKERAY, REVISED AND CONSIDERABLY ENLARGED.

Contains two unpublished letters of A. C. Swinburne, Thackeray's contributions to "The National Standard," "The Snob," also "Dickens in France," "Letters on the Fine Arts," "Elizabeth Brownrigge: A Tale," &c.

"Thackeray collectors, however, have only to be told that NONE OF THE PIECES NOW PRINTED APPEAR IN THE TWO VOLUMES RECENTLY ISSUED by Messrs Smith, Elder, & Co., in order to make them desire their possession. They will also welcome the revision of the Bibliography, since it now presents a complete list, arranged in chronological order, of Thackeray's published writings in prose and verse, and also of his sketches and drawings." —*Daily Chronicle.*

"'Sultan Stork' . . . . is undoubtedly the work of Mr Thackeray, and is quite pretty and funny enough to have found a place in his collected miscellanies. 'Dickens in France' is as good in its way as Mr Thackeray's analysis of Alexander Dumas' 'Kean' in the 'Paris Sketch-Book.' . . . There are other slight sketches in this volume which are evidently by Mr Thackeray, and several of his *obiter dicta* in them are worth preserving. . . . We do not assume to fix Mr Thackeray's rank or to appraise his merits as an art critic. We only know that, in our opinion, few of his minor writings are so pleasant to read as his shrewd and genial comments on modern painters and paintings."—*Saturday Review.*

"ADMIRERS OF THACKERAY MAY BE GRATEFUL FOR A REPRINT OF 'SULTAN STORK.'"—*Athenæum.*

Demy 8vo, pp. viii. and 68, Parchment, 7s. 6d.

Primitive Symbolism as Illustrated in Phallic Worship;

Or, The Reproductive Principle.

By HODDER M. WESTROPP.

With an Introduction by General Forlong.

"This work is a *multum in parvo* of the growth and spread of Phallicism, as we commonly call the worship of nature or fertilizing powers. I felt, when solicited to enlarge and illustrate it on the sudden death of the lamented author, that it would be desecration to touch so complete a compendium by ONE OF THE MOST COMPETENT AND SOUNDEST THINKERS WHO HAVE WRITTEN ON THIS WORLD-WIDE FAITH. None knew better or saw more clearly than Mr Westropp that in this oldest symbolism and worship lay the foundations of all the goodly systems we call Religions."—J. G. R. FORLONG.

"A well-selected repertory of facts illustrating this subject, which should be read by all who are interested in the study of the growth of religions."—*Westminster Review*.

Fcap. 8vo, 80 pp., Vellum, 10s. 6d.

Beauty and the Beast;

Or, a Rough Outside with a Gentle Heart.

A Poem.

By CHARLES LAMB.

Now first Reprinted from the Original Edition of 1811, with Preface and Notes by Richard Herne Shepherd.

For three quarters of a century this charming fragment of Lamb's genius lay buried; even the author seems to have forgotten its existence, since we find no reference, either direct or indirect, to the little tale in Lamb's published correspondence, or in any of the Lamb books. The credit of a discovery highly interesting to all lovers of Charles Lamb is due to the industry and sagacity of Mr John Pearson, formerly of 15 York Street, Covent Garden.

The publisher has now endeavoured to place the booklet beyond future chance of loss by reproducing ONE HUNDRED COPIES for the use of libraries and collectors.

18mo, pp. xxvi. and 174, Cloth extra, 2s.

Wellerisms,

From "Pickwick" and "Master Humphrey's Clock."

SELECTED BY CHARLES F. RIDEAL,
AND EDITED, WITH AN INTRODUCTION, BY CHARLES KENT.

Among the Contents are :—Sam Weller's Introduction—Old Weller at Doctor's Commons—Sam on a Legal Case—Self-acting Ink—Out with It—Sam's Old White Hat—Independent Voters—Proud o' the Title—The Weller Philosophy—The Twopenny Rope—Job Trotter's Tears—Sam's Misgivings as to Mr Pickwick—Clear the Way for the Wheelbarrow—Unpacking the Lunch Hamper—Battledore and Shuttlecock—A True Londoner—Spoiling the Beadle—Old Weller's Remedy for the Gout—Sam on Cabs—Poverty and Oysters—Old Weller on Pikes—Sam's Power of Suction—Veller and Gammon—Sam as Master of the Ceremonies—Sam before Mr Nupkins—Sam's Introduction to Mary and the Cook—Something behind the Door—Sam and Master Bardell—Good Wishes to Messrs Dodson & Fogg—Sam and his Mother-in-Law—The Shepherd's Water Rates—Stiggins as an Arithmetician—Sam and the Fat Boy—Compact and Comfortable—Apologue of the Fat Man's Watch—Medical Students—Sam Subpœnaed—Disappearance of the "Sausage" Maker—Sam Weller's Valentine—Old Weller's Plot—Tea Drinking at Brick Lane—The Soldier's Evidence Inadmissible—Sam's "Wision" Limited—A Friendly "Swarry"—The Killebeate—Sam and the Surly Groom—Mr Pickwick's Dark Lantern—The Little Dirty-faced Man—Old Weller Inexorable—Away with Melancholy—Post Boys and Donkeys—A Vessel—Old Weller's Threat—Sam's Dismissal of the Fat Boy—Is she a "Widder"?—Bill Blinder's Request—The Watch-box Boy.

" . . . . THE BEST SAYINGS of the immortal Sam and his sportive parent are collected here. The book may be taken up for a few minutes with the certainty of affording amusement, and it can be carried away in the pocket." —*Literary World.*

"It was a very good idea . . . the extracts are very numerous . . . here nothing is missed."—*Glasgow Herald.*

Demy 8vo, pp. 99, with Protractor and 16 plates, coloured and plain. Cloth gilt, 7s. 6d.

Geometrical Psychology;

Or, The Science of Representation.

AN ABSTRACT OF THE THEORIES AND DIAGRAMS OF B. W. BETTS.

BY LOUISA S. COOK.

"His attempt seems to have taken a similar direction to that of George Boole in logic, with the difference that, whereas Boole's expression of the Laws of Thought is algebraic, Betts' expresses mind-growth geometrically;

that is to say, his growth-formulæ are expressed in numerical series, of which each can be pictured to the eye in a corresponding curve. When the series are thus represented, they are found to resemble the forms of leaves and flowers."—*Mary Boole, in " Symbolic Methods of Study."*

The *Pall Mall Gazette*, in a characteristic article entitled, "Very Methodical Madness," allows that "Like Rosicrucianism, esoteric Buddhism, and other forms of the mystically incomprehensible, it seems to exercise a magnetic influence upon many minds by no means as foolish as its original inventor's."

"This work is the result of more than twenty years' application to the discovery of a method of representing human consciousness in its various stages of development by means of geometrical figures—it is, in fact, THE APPLICATION OF MATHEMATICAL SYMBOLOGY TO METAPHYSICS. This idea will be new to many of our readers; indeed, so far as we know, Mr Betts is the only man who has tried to work out a coherent system of this kind, though his work unfortunately remains imperfect."—*Theosophist*, June 1887.

8vo, pp. 32, Wrapper, 1s.

On Mesmerism.

By A. P. SINNETT.

Issued as a *Transaction* of the London Lodge of the Theosophical Society, of which Mr Sinnett is President, this pamphlet forms AN ADMIRABLE INTRODUCTION to the study of Mesmerism.

LONDON: GEORGE REDWAY.

www.ingramcontent.com/pod-product-compliance
Lightning Source LLC
Chambersburg PA
CBHW021731220426
43662CB00008B/796